Science as Cognitive Process

Robert A. Rubinstein

Charles D. Laughlin, Jr.

John McManus

University of Pennsylvania Press
Philadelphia

Science as
Cognitive
Process

Toward an Empirical Philosophy of Science

Copyright © 1984 by the University of Pennsylvania Press.
All rights reserved

Library of Congress Cataloging in Publication Data
Rubinstein, Robert A.
　Science as cognitive process.

　Bibliography: p.
　Includes index.
　1. Science—Philosophy. 2. Science—Psychological
aspects. 3. Science—Social aspects. I. Laughlin,
Charles D., 1938–   . II. McManus, John, 1942–
III. Title.
Q175.R78 1984   501   83–14730
ISBN 0-8122-7911-5

Printed in the United States of America

This book is dedicated with affection to **Earl W. Count**

Science is not merely a collection of facts and formulas. It is pre-eminently a way of dealing with experience. The word may be appropriately used as a verb; one *sciences*.

> —Leslie A. White, 1938
> "Science Is *Sciencing*," *Philosophy of Science* 5:369–89.

Contents

Preface   xiii

Introduction   xvii

One
The Anthropological Study of Cognition   1
    Cognitive Anthropology Briefly Characterized   1
    Cognitive Anthropology Critically Evaluated   7

Two
Process and Structure in Cognition   21
    The Cognized and Operational Environments   21
    Operational and Cognized Logics   31

Three
Multivariable Cognitive Development   37
    The Concept of Structure   37
    The Concept of Development   40
    Structural Developmental Variables   41
    Implications for Social Science Research   51

Four
Cognition and Paradigmatic Science   61
    The Nature of Paradigms   62
    A Multilevel View of Paradigmatic Science   64
    Paradigms and the Issue of Anomalies   67
    Some Proposed Principles   78

Five
Theory Reduction as Cognitive Process 85
   Anthropology and the Philosophy of Science   86
   The Received View of Theory Reduction   88
   The Structural Basis of Theory Reduction   89
   Reduction by Incorporation   96

Six
Explanation and Cognition 101
   Anthropology and Explanation   101
   Critiques of D-N and I-S Explanation   107
   The Problem of Explanation in a
       Cognitive Process Approach to Science   113
   Statistical-Relevance Explanation   115
   The S-R Model in Anthropological Research:
       An Archaeological Case   119

Seven
Cognition and the Justification of Induction 123
   Induction and Cognition   124
   Justification of Induction   128
   Induction in the Context of Science   130
   Science in Perspective   136

Eight
Experience and the Internal Image in Scientific Thinking 141
   Kinds of Knowing   142
   Forgotten Levels of Knowledge   144
   Two Operational Environments   145
   Where the Data Hide   147
   Clues from Clinical Psychology   151
   Entering Thought at the Image Level   153
   The Advantages of Images   155
   Knowledge and Transformations   158

Bibliography   161

Author Index   177

Subject Index   181

Tables

Table 1.1 Evidence for a Dichotomy Between Comprehension and Production in Linguistic Theory   19
Table 2.1 Some Relations of Isomorphism in E_c/E_o Processes   26
Table 7.1 Features of General Cognition and Features of Sciencing Compared   137

Figures

Figure 1.1 Cognitive Maps for a Simple Social Interaction System   15–16
Figure 2.1 Levels of Structure Operating in the Human Organism as an Actor in a Social System   29
Figure 2.2 General Relationship Between Environmental and Cognitive Complexity   32
Figure 3.1 Lower Levels of Conceptual Structuring   48
Figure 3.2 Higher Levels of Conceptual Structuring   49
Figure 7.1 Induction and Deduction as Aspects of Cognitive Adaptation   127
Figure 7.2 Inductive/Deductive Alternation in Science   135

Preface

The collaboration from which this book results began in late 1972. At that time Charles Laughlin and Eugene d'Aquili were developing the first formal presentation of biogenetic structural theory. Their work was accompanied by a great deal of excitement, and there gathered around them a number of colleagues who shared the promise that the newly developed theory held. Each of us knew that the publication of *Biogenetic Structuralism* (New York: Columbia University Press, 1974) marked the beginning of a research tradition rather than the summary systematization of a body of knowledge about human social life. The informally constituted "biogenetic structuralism research group" undertook, separately and in various collaborative arrangements, to work out some of the applications and implications of the theory.

In order to explore more fully the idea that science might usefully be studied in the same way as other aspects of human social life, I wrote in 1973 and 1974 several working papers examining scientific explanation, inductive logic in scientific reasoning, theory reduction, and paradigmatic science from the biogenetic structural point of view. While preparing for the symposium "Studies in Biogenetic Structuralism," held at the 1975 annual meeting of the American Anthropological Association, Charlie Laughlin and I worked together to revise and further develop the working paper on theory reduction (and we subsequently published it in *Current Anthropology*, 1977). At about that time we agreed that an integrated study of science from this perspective would both add a necessary methodological undergirding to biogenetic structural research and provide interesting answers to some unresolved questions in the philosophy of science, and we asked John McManus to join us in such a project.

Successful scholarly collaboration is often difficult, and the course of our work on this book has been made more so by its having been interrupted several times because one or another of us was abroad for an extended period. Despite this, and our having to carry out our joint work by mail because of the distances between us since 1977, the collaboration has been rewarding and instructive.

The analysis of science offered in this book marks a point in an ongoing process of inquiry. As such it suggests the outlines of a research program in the empirical philosophy of science and in anthropology. We have tried to make clear what we think are some of the major directions this research program might take, and what some of the answers it may yield will look like. We hope that this book will lead others of our colleagues to join in the project of studying critically the nature of science as a part of social life.

Earlier formulations of some of the material in this book have been published elsewhere, particularly by Rubinstein and Laughlin in *Current Anthropology* (Vol. 18, 1977), by Laughlin and Brady in *Extinction and Survival in Human Populations* (New York: Columbia University Press, 1978), by Laughlin and McManus in *The Spectrum of Ritual* (New York: Columbia University Press, 1979), and by Rubinstein in *Human Relations* (Vol. 34, 1981).

During the time we worked on this book, we were helped by many friends and colleagues. For critical discussions of this material we thank Michael Agar, Jerrold Aronson, John H. Beatty, Aarron Ben-Zeev, James P. Boggs, Ivan A. Brady, Rodney Byrne, Richard P. Chaney, Gary B. Coombs, John Cove, Eugene d'Aquili, Roy S. Dickens, Bruce Donaldson, Lawrence E. Fisher, Brian Foster, Mary LeCron Foster, Larissa Lomnitz, Fred Plog,

Iain Prattis, Arie de Ruijter, Jon Shearer, Christopher D. Stephens, Oswald Werner, and Ina Jane Wundram.

Robert L. Blakely, Ronald Cohen, David Hufford, Susan Mills Isen, Rik Pinxten, Robert L. Welsch, and William C. Wimsatt each read an earlier draft of this book and made helpful comments and criticisms, for which we are very grateful.

We especially appreciate the help of H. Stephen Straight and of Sol Tax, each of whom not only read and commented on earlier drafts of this book, but also offered at several critical points their support and encouragement for this project.

Robert A. Rubinstein

Chicago, Illinois
May 1983

Introduction

Most people can agree easily enough on a definition of a particular scientific discipline. A scientific discipline, they might say, is the systematic study of a particular class of phenomena. For example, biology is the systematic study of living organisms and their vital processes, and psychology is the study of mind and behavior. An equally likely answer might be that a scientific discipline is the activity of a group of people who recognize themselves and are recognized by others as belonging to that discipline, and that this group of people undertake the systematic study of a particular class of phenomena. Thus, biology is what people who call themselves biologists, and are called biologists by others, do when they study living organisms and their vital processes.

It is much harder to agree on a workable definition of science. Answers that parallel those describing particular scientific disciplines are problematic. To define science as the systematic study of phenomena is to leave both "systematic study" and "phenomena" undefined. And, distinctions between the kinds of study included within science are quickly drawn. People talk routinely about "hard" versus "soft" science, "experimental" versus "theoretical" science, or "natural" versus "social" science, for example. Once these sorts of distinctions are drawn, one or another discipline, or set of disciplines, is sometimes said to be more or less scientific than others, or to be more or less systematic, or to be devoted to the study of more or less well defined classes of phenomena. Agreeing on a listing of what sorts of people are properly called scientists turns out not to be a simple task.

Certainly there will be some, like physicists, who are pretty much on everyone's list. But there is an equal number of candidates for such a list that would spark disagreement. Behavioral psychologists, for instance, might want to exclude transpersonal psychologists from such a list, and some physicists might want to exclude all anthropologists.

Before these kinds of disputes can be settled, we need to have answers to an intertwined set of questions about what it is that gives science its special character. That is, we need to be relatively well satisfied that we know what the nature of the scientific enterprise is. The ready response is that science is characterized by the application of the scientific method of specifying a problem, collecting data through observation or experiment, and formulating and testing hypotheses. While this response may help differentiate in a general way some activities that we call science from some we would not, it only suggests the broadest outlines of an adequate response.

For more detailed responses, we traditionally have asked practicing scientists to reflect on their experiences, and we have addressed the question also to those philosophers who take science as their special area of concern. Increasingly over the past three decades the question "What is the nature of science?" has been asked more often of philosophers of science than of reflective scientists. During that time the philosophy of science grew rapidly, and the outlines of a coherent philosophical analysis of science were set out. Generally acceptable accounts of scientific explanation, the nature and role of scientific theories, theory reduction in science, and other aspects of science were elaborated. For a while, at least, it looked as though the major obstacles to an understanding of science had yielded to philosophical analysis.

To those working outside the philosophy of science, this happy appearance lasted well into the 1970s, and some practicing scientists even tried to fit their own work to the model of science presented by philosophers. (For examples, see Fritz and Plog's [1970] work in archaeology, Jarvie's [1967] general statements about methodology in social anthropology and his treatment of cargo cults as an anthropological problem, and Skinner's [1953] program for behavioristic psychology.) In fact, however, the wide-ranging agreement about how to approach the understanding of science that had characterized the philosophy of science at the beginning of the 1960s had within a decade begun to dissolve under the critical scrutiny of philosophers themselves. For instance, Reichenbach's (1938) view that philosophers should confine their inquiry into the nature of science to the examination of the context of justification (that portion of the scientific enterprise where well-formed theoretical propositions are evaluated) and leave the study of the context of discovery (that portion of the scientific enterprise wherein theoretical propositions are formulated) to others was challenged in major works by then younger philosophers of science like Hanson (1958) and Kuhn (1962).

The focus of these challenges was essentially that by excluding from consideration empirical data about the progress of science, the received view offered analyses of the nature of science that applied only in principle (see Suppe 1977; Boyd 1972; Wimsatt 1979). In practice the behavior of scientists—even in the context of justification—did not appear to fit well the descriptive and normative offering of the received philosophical account of science.

As a result of these challenges the inadequacies of the received account of science began to be more widely known, and empirical considerations began to be more common in the work of philosophers of science. Principally that empirical material came from the history, and to a lesser extent from the sociology, of science. A number of alternative accounts of science, or of particular aspects of science, have resulted, though none has achieved the widespread acceptance that the earlier received view had gained (Suppe 1977).

It was this lack of consensus that we found when we began to explore together the nature of the scientific enterprise. The then relatively recent attempts to combine philosophical analyses with historical and sociological data in the service of achieving an adequate understanding of the nature of science were at once welcome and puzzling. We welcomed them because as students of human behavior we saw science as a form of human activity best approached in the manner that might be used to gain an understanding of other forms of human behavior such as the study of social organization, or of communicative competence to which historical and sociological data are clearly relevant. We were puzzled, though, because we had come to believe that any adequate account of human behavior would combine data from many different levels of human activity, not only historical and sociological, but also biological, psychological, and cultural (cf. Bateson 1941; Count 1973). While philosophers had finally overcome their reluctance to include historical and sociological data in their deliberations, questions of psychology and culture were still for the most part either viewed as illegitimate or resolved by reference to introspection.

This was unfortunate because anthropologists and psychologists had shown that answers to questions of psychology and culture could reveal a great deal about the nature of human behavior. That work had shown too that introspection, especially when used alone as a method of investigation, was notoriously inadequate for answering such questions. That introspection or exclusion were the predominant ways of dealing with psychology and culture presented, in our opinion, a central difficulty for the philosophical analysis of science. This book results from our attempt to correct this situation by analyzing science from the perspective of anthropology and cross-cultural psychology.

In our view, science extends in a specific formalized normative system structures and processes that are basic to human cognition and behavior in general. To give an adequate account of science is to describe an activity, a set of processes we call *sciencing*.

Seen in this way, a prerequisite for the analysis of science is the description of the structure and operation of functioning cognitive systems in general. The task is in major part to reveal in a systematic fashion the way in which human cognitive systems interact with their environments and to see what implications this holds for an understanding of the specific processes of sciencing.

The first step in such an investigation is to identify the general structures and processes of cognition. When we began we thought this first step straightforward. Anthropology and psychology each had rich research literatures exploring various aspects of cognition and culture, and it was to those literatures that we turned in our initial attempt to identify the general processes of cognition relevant to the study of sciencing.

Despite claims to the contrary, it turned out that these research traditions focused not on describing and accounting for the ongoing performance of cognitive tasks but on the formal nature of cognitive structures. Frequently, especially in anthropology, they treated the content of thought as primary, and the study of both structure and process was generally neglected (see Cole and Scribner 1975). In short, if the goal of cognitive science is taken to be the modeling of functioning, active cognitive systems, then the existing models and methods fell short of yielding adequate analyses. This led us to develop the extended critique of cognitive studies in anthropology presented in Chapter 1.

We sought then to develop an alternative model of human cognition and behavior able to guide the study of performing cognitive systems. This led us to refine and extend a theoretical position which was presented by Laughlin and d'Aquili (1974) and which, after some further development, had been applied fruitfully to the study of ritual (d'Aquili et al. 1979), adaptation to changing environments (Laughlin and Brady 1978), and language acquisition (Rubinstein 1976, 1979a). This model, and its revisions, incorporate elements from a number of disciplines, mainly from Piagetian and developmental psychology, social and cultural anthropology, evolutionary biology, and information-processing theory. The approach focuses on discovering the processes people use to keep their views and understandings of how the world is in some passably accurate relationship to how the world actually is. From this model we selected material that seemed particularly appropriate for the analysis of sciencing to use as a base from which to begin this inquiry. The concepts and data from this material are presented in Chapter 2.

The model suggests that the development of cognitive structures and processes occurs simultaneously in several domains of thought. These developing structures and processes interact to facilitate further development. But development in these various cognitive and affective domains depends upon prior development in other, more basic, cognitive capacities. Chapter 3 reviews several theories of the growth of cognitive structure and process. It then draws implications for the conduct of science from this material. Principally, the chapter reviews the case for careful attention to differential development in various cognitive domains.

The argument advanced is that the success of science depends upon clear and continuous communication. That clarity of communication among scientists (and in the case of those sciences that take people as their subject, between scientists and the people they study) depends upon how exchanged information is structured. Information presented in a manner more highly structured than the inquiring scientist is prepared to cope with may well be misunderstood. Chapter 3 reviews the systematic ways in which such misunderstandings might arise and the systematic biases that may result from too little attention to the relation of environmental complexity to the complexity of the cognitive systems of the working scientist (cf. Bateson 1935; Wimsatt 1980a).

Chapter 4 continues with the exploration of how cognitive process and structure affect sciencing as a social phenomenon. The principal vehicle for this analysis is the notion of paradigmatic science introduced by Kuhn (1962). In the chapter we argue that only by explicitly introducing considerations of cognitive and perceptual development, and seeing these processes as relevant to the study of sciencing, can an adequate understanding of the operation of paradigms be developed. It then shows how analyses of transformations in cognitive structures and processes can help account for the distinction between "seeing" and "seeing that" in science proposed by Hanson (1958), and for the apparent ability of scientists to set aside important anomalies between data and theory.

One of the major reasons for the widespread acceptance of the received philosophical view of science was its apparent success in giving a coherent descriptive and normative account of scientific explanation. Basically the view is one of explanation as argument, wherein the phenomenon being explained is made to appear likely to have occurred (Hempel 1965). The form of the explanation was by deductive derivation via subsumption under general laws (or by inductive support via subsumption under statistical generalizations). Such explanation was extended to a treatment of theory reduction in science, which was then viewed as a special case of explanation by subsumption under laws (Nagel 1961). It turned out that neither in the special case of theory reduction, where difficulties first became evident, nor in the general case of the explanation of empirical events was the account problem-free.

Chapters 5 and 6 take up theory reduction and explanation from the view that both are methods of increasing our understanding of the world, rather than exercises in the elaboration of formal systems. On this view the processes underlying the growth, development, and transformation of cognitive structures and processes ought to be preserved as features of scientific understanding. The result of such a view is the specification of a set of constraints on the nature of explanation and theory reduction which must be observed in any adequate descriptive or normative account. (Ideally, in fact, the normative and descriptive accounts should be very similar, and perhaps even identical.) In Chapters 5 and 6 we set out what we think are the important empirical constraints on satisfactory accounts of explanation and theory reduction and show how these are consistent with some, but not other, extant philosophical accounts of these issues.

In coming to understand the nature of sciencing, it is necessary that we have some way of responding to Hume's challenge to the use of scientific induction as an epistemological tool. Lacking outright justification of scientific induction, it is at least important to understand why it appears central in Western epistemological thinking and to evaluate it in relation to other epistemological tools. Chapter 7 proposes that scientific induction is the normative expression of just one portion of the process of acquiring an understanding of the world and thus of scientific model-building. We conclude that scientists should pay closer attention to anomalous information and to data from multiple levels of analysis. This requires that scientists pay closer attention to seeing themselves as important "instruments" of scientific inquiry and that they expand their conception of their role in sciencing.

The final chapter of the book takes up this theme and argues that, in the service of becoming more sensitive to their role in the process of sciencing, scientists must recognize the complementary roles played by figurative and cognitive sources of knowledge in the process of developing sophisticated and powerful models of the world. In Chapter 8 we argue that by making it normatively important for scientists to be aware of figurative as well as of other sources of information, the complexity and adequacy of science will be increased.

This book presents a first, preliminary view of science as cognitive process. Our intention is to open this approach as a legitimate area of discourse, not to offer a final account of sciencing. Indeed, if our model of sciencing is at all worthy, the view we present here will need to be refined and tested on a number of different levels before it can be satisfactory. Moreover, the analysis presented here is not put forward as a comprehensive treatment of all issues that arise in considering science as sciencing. Still to be explored from this cognitive perspective, for example, is the adequacy of various interpretations of the nature of scientific theory. Our hope for this book is merely that it open discussion of hitherto neglected aspects of science as cognitive process.

One

The Anthropological Study of Cognition

Science is a function of human cognition. Our concern is with how the cognitive system works, how it succeeds in representing, or in modeling, the environment. Hence, we begin with the premise that a complete understanding of science necessarily includes the investigation of human cognition, at least insofar as cognitive processes underlie scientific inquiry.

Within anthropology there is a long-standing research tradition focusing on the cross-cultural analysis of human cognition. This tradition, often called cognitive anthropology or ethnoscience, provides sophisticated investigative techniques and useful initial assumptions for the study of human behavior and cognition. It is, however, inadequate for the investigation of science. In this chapter we show why. To do this we characterize cognitive anthropology and offer an account of its shortcomings for our present purposes. Chapter 2 then presents a discussion of some of the key concepts in an alternative approach to the study of human cognition and behavior. We emphasize that our aim here is not to disparage the theoretical and empirical accomplishments of cognitive anthropology, but to lay the groundwork for the construction of an alternative view of the processes of human cognition and of science.

Cognitive Anthropology Briefly Characterized

Richard Chaney (1972; Chaney, Morton, and Moore 1972) has described for specific cases of anthropological research how the conceptual systems underlying observation have profound effects upon the results of research programs. The following characterization of cognitive anthropology provides some insight into its particular outlook, heuristics, and biases. Before undertaking this review, we offer the following caveat. Theoretical approaches in science are not distinct monolithic sets of propositions. It is more often the case that theoretical approaches overlap and that distinctions between approaches blur. As a result, even a long and detailed discussion that attempts to characterize a specific theoretical approach in isolation from the larger concerns of the discipline or research program risks distorting through caricature the outlines and details of that particular approach. Our view of cognitive anthropology should not be seen as a general critique, which could be achieved only through discussion of its internal complexity and its relation to anthropology as a whole. Instead, the present exposition is a *specific* critique of various aspects of cognitive anthropology as they relate to the study of science.

Perhaps the best way to characterize cognitive anthropology is to look at what cognitive anthropologists say about what they do. Conklin (1964:93; see also Hammel 1965) notes:

> An adequate ethnography is . . . considered to include the culturally significant arrangement of productive statements about the relationships obtaining among locally defined categories and contexts . . . within a given social matrix.

And Tyler (1969:3) stresses that cognitive anthropology

> focuses on *discovering* how different peoples organize and use their culture. This is not so much a search for some generalized unit of behavioral analysis as it is an attempt to understand the *organizing principles underlying behavior*. It is assumed that each people has a unique system of perceiving and orga-

nizing material phenomena—things, events, behavior, and emotions. (Emphasis in the original.)

Similarly, Frake (1962:28–30) argues that the ethnographer's task is not to get the words for things but to find the things that go with words. Because not even the most concrete physical objects can be identified apart from a culturally defined conceptual system, ethnographers should strive to define objects according to their informant's conceptual system. Their task is not one of linguistic recording; it is one of identifying objects and processes in their informants' environments.

Cognitive anthropology focuses on the explication of intragroup phenomena. Typically, these are specific to the cultural group studied, with little or no attention given to the question of how they may relate to possible universal principles for the organization of human behavior. Cognitive anthropology attempts to understand how members of a culture go about modeling features of their environment. Cognitive methods seek to provide analyses that reveal how to view the world through the informants' eyes, to, as it were, "get into the informants' mind." It is important to note that cognitive anthropological analyses are not conceived simply as descriptions of a series of behavioral regularities that satisfactorily summarize the informant's actions. Rather, these analyses are said to display the informants' principles of cognitive organization and to explain how these principles are employed in perception, thought, and action. Cognitive anthropologists thus claim to be particularistic, constructivistic, and realistic in their orientation to the study of mind.

Another approach is to look at the kinds of things cognitive anthropology takes as proper objects of study. Cognitive anthropologists seek to discover formal realizations of the mental entities under study, rather than to infer such mental entities from their data. What cognitive anthropologists seek to discover, to identify, to explicate, and to formalize is the *structure* that underlies behavioral forms. Often this search focuses on linguistic structures (the structures that underlie linguistic forms), although it also deals at times with behavioral analogs (or homologs) of those structures.

In fact, the main concern of cognitive anthropology is with the discovery of structure in the sense set out by Piaget (1970:3–16). It is only because "structure exists apart from the theoretician" that it can be discovered. Furthermore, like those of Piaget, the structures investigated by cognitive anthropologists are thought to form wholes, to have rules governing the composition of structure (rules of transformation), and to be self-regulating.

Not only do cognitive anthropologists seek to discover structure, they increasingly seek also to formalize their analyses (see Werner and Fenton 1973). Formal analysis places "its emphasis on internal consistency, completeness and form" (Tyler 1969:13) and consists of the specification of a set of primitive elements, a set of rules for operating on these, and the application of these rules to a specific body of data in a way that shows them to be expected as consequences of underlying principles presumed to be at their source (Lounsbury 1964). A formal analysis is said to be complete when it describes the relations among all the elements comprising the domain under study.

Since cognitive anthropologists seek to discover structure and then attempt to formalize those structures with varying degrees of rigor, it may be characterized as a form of structuralism focusing on intragroup phenomena. Implicit in this structuralist program is the view that human cognition and behavior is rule-following (or at least rule-oriented), in the sense that behavior is produced by the application of a set of rules or principles to specific situations. Cognitive anthropology is concerned with the rules that people use for organizing and manipulating their culture.

A subsidiary proposition of this assumption is that the study of social behavior is the study of symbolic systems. Moreover, what is of interest is the structure of the system itself rather than how the system is used. This investigation often boils down to the formal semantic analysis of a lexical domain, though the behavioral-cognitive domain being studied may not be linguistic. Although the study of nonlinguistic domains usually adds a unique set of problems (such as how to define the usually more vague and less well bounded behavioral analogs of lexical domains), the theoretical assumptions remain in principle unchanged (see Werner and Fenton 1973; compare Quine 1960, 1969).

Cognitive anthropology also makes the assumption that human cognition and behavior are hierarchically organized on the basis of sets of distinctive features. This assumption shows most clearly in the analytic tools that cognitive anthropology employs. One of the most popular among these is taxonomic analysis (Spradley 1970; Werner and Fenton 1973; Lancy and Strathern 1981). The development of a taxonomy involves the explication of contrast between distinctive features among and between levels of organization, and lower levels are necessarily included in higher ones.

It might be objected that these conclusions are simply the automatic results of the form of analysis employed. In response, we point out that methodological techniques are closely related to the theoretical programs that utilize them. Method is logically implied by theory (Suppe 1977; Wimsatt 1980a). The analysis of behavior and cognition into hierarchically ordered levels reflects implicit assumptions about these phenomena.

Another set of interesting assumptions made by cognitive anthropology also stems from the central use of formal analysis. These are (1) that analyses should be presented with a maximum amount of specificity and clarity, (2) that an analysis should be exhaustive—it should specify all the relationships between all elements in a domain, and (3) that a formal account should reduce ambiguity by removing redundant relationships among all elements. The first assumption is straightforward.

Precision is a desideratum of all scientific explanation (Hempel 1965; Nagel 1961; Salmon 1971). The second assumption too poses no special problems; it is surely desirable for an account of some phenomenon to be as complete as possible. However, the assumption that an analysis should be as simple as possible is problematic. It suggests that the simpler the account the better. In the following section in this chapter, and later in the book, we argue that this maxim is not necessarily desirable in the conduct of science in particular, or even in the study of human cognition and behavior in general. We suggest later that the pursuit of simple models of a complex reality is detrimental to the development of accurate knowledge of the environment.

Cognitive anthropology, then, makes three sets of assumptions about the nature of human cognition and behavior. Central to the first set is the notion that human behavior is rule-following. This raises questions about the nature of these rules, their ontological status, the relation of these rules to individual actors, and the like (Straight 1974). The second set focuses on the notion that human cognition and behavior should be seen as hierarchically ordered, suggesting questions about the uniqueness of an individual rule system, the compatibility of rule systems, the verification of such hierarchical ordering, and the relationship of behavior to individual rule systems. The final set of assumptions incorporates the view that the simpler these rules the better, thus raising the question Are rules used by informants always the simplest, or is it possible for complex and intricate sets of rules to provide better adaptive fit with the environment?

The next section in this chapter discusses these and other related questions. This discussion shows that cognitive anthropology as presently constituted is fundamentally incapable of arriving at a theory of cognitive process, and thus is incapable of supporting an analysis of science.

Cognitive Anthropology Critically Evaluated

A discussion of the shortcomings of cognitive anthropology can be organized around four main issues: psychological reality; the sharedness of cognitive systems; the competence versus performance dichotomy; and explanatory perspective in anthropological analysis.

Psychological Reality

To claim that a particular analysis of human behavior has psychological reality is to assert at least two things. First, one is claiming that the analysis explicates the cognitive processes that were involved in the formulation of the analyzed behavior. Put another way, the claim is that, at a minimum, the ethnographer has identified the significant stimuli in the informants' environment as well as the principles of organization that the informants use to model these stimuli and to arrive at responses to them. The importance of this is that the ethnographer claims, in the language of cognitive anthropology, to have gotten inside the informant's mind by inferring the particular set of cognitive activities that went into the production of particular behaviors. The second claim is that, given knowledge of the situational constraints, the analysis will allow the ethnographer (or other person) to produce appropriate, acceptable behavior in the informant's culture.

If an ethnographer undertakes to discover significant stimuli, and to explicate the principles whereby those stimuli are organized in the planning and execution of action, then she or he is obliged to claim psychological reality for the analysis. Cognitive anthropological analyses typically claim to present the calculus used by informants, *rather than some operationally equivalent calculus.* Since there are always alternative calculi that are equally effective in accounting for the observed behavior (Hunter 1977:39; Wallace 1970:87–88), how can we tell that we have identified the one used by the informants? Four characteristic ways of "resolving" the problem of psychological reality emerge from a review of the literature: (1) The question is dismissed as unimportant; (2) an appeal is made to "simplicity," to some way of comparing the formal elegance of the analysis to that of its operationally equivalent rivals; (3) introspective confirmation is sought; and (4) the success of the analysis as an account of the observed data is taken as sufficient evidence of its psychological reality.

To set the question of psychological reality aside by saying that it is trivial and unimportant is to abandon the central purpose of cognitive analysis which is to derive a culturally sensitive (particularistic), experience-based (constructivistic) model of real mental processes. In addition, the response that a model is adequate if and only if it predicts or retrodicts to decisions or actions fails because we are concerned with understanding cognition itself as an element in the behavioral environment. If we want to explore the interplay between active cognitive systems and their environments, this response to the question of psychological reality is untenable.

The second "resolution" suggests that the formally more parsimonious analysis is the psychologically real one. The dictum typically is that, all else being equal, the briefer the analysis the better. This is an unsatisfactory response for three reasons. First, the consistent application of the principle of parsimony would require that, with the development of an analysis that is more parsimonious than an analysis earlier accepted as psychologically real, we replace the earlier one with the newly derived one. This would create serious problems, for if the first psychologically real calculus is removed in favor of the second, we in effect deny its reality. Here it would be possible to hedge a bit and to modify the criterion of psychological reality: All else being equal, the shorter the analysis the *more* psychologically real it is. But this is a considerably different claim from the one which says that the ethnographer can discover the psychologically real calculus. In fact, this would force us to relinquish any claim for the cognitive status of the analysis. Thus, for practical reasons, parsimony cannot be used as a necessary condition for determining psychological reality.

Furthermore, empirical evidence suggests that cognitive systems and strategies are not always the most streamlined. Indeed, it appears that they often turn out to be surprisingly clumsy and indirect. As we note in the following chapter, what is parsimonious for individuals at one level of cognitive functioning may be horrendously complex for individuals functioning at another level. The use of parsimony provides no help in resolving the problem of psychological reality. This failure parallels the problems involved in the application of the principle of parsimony to scientific theories in general. How, for instance, is the criterion for specifying the degree of parsimony achieved to be formulated? Quine (1964) has argued that the application of the principle of parsimony may actually confound rather than facilitate the development of sound scientific theory.

The third attempted resolution of the psychological reality question involves recourse to the informants' introspective confirmation of the ethnographer's analysis. This approach relies on the informants' ability to describe their internal states to the ethnographer. In effect, this method of resolution involves the presentation of the analysis to the informant with the question What do you think, is this the way you go about it? (Even though this is a caricature of the approach, the problems facing this formulation remain intact in more sophisticated versions of this method.) The objection to this method is: How can an ethnographer know the introspection is accurate?

Even if every informant could provide accurate introspective accounts of his or her cognitive processes, the problem stands, for how is it possible for an ethnographer to know that although the informants are capable of accurate introspections they are willing to relate these? Had Turnbull (1972, 1978) attempted to confirm his analysis of the Ik via introspective techniques, he certainly would have met with little success. On the other end of the spectrum, in a society where interaction with the ethnographer might provide some tangible or other reward for an informant, informants might "feed" the ethnographer what they think is wanted, thus giving rise to spurious results (Chagnon 1977). This discussion has an analog in experimental psychology, where it is labeled "experimenter bias" (Rosenthal 1963). It is possible that the information which informants give a researcher may be biased by any one of a number of aspects of the interaction setting.

It is clear that introspection does not offer a solution to the problem of psychological reality in cognitive analysis. Human psychology simply cannot be based on introspection, since all people can do is introspect to the content of their cognitive systems, while action may involve uncognized internal cognitive processes and unconscious structures underlying those processes. However, our assertion that introspection is not a solution to the problem of psychological reality does not imply that introspective accounts from informants have no place in cognitive-anthropological or other research. In fact, we think that introspective accounts from informants *and* from anthropologists have a large and important role to play in the study of human cognition and behavior, but as further data to be considered rather than as privileged sources of insight into the mind (Laughlin and d'Aquili 1974; Rubinstein 1976; Laughlin and Brady 1978).

The final response to the psychological reality question may be characterized as follows: The ethnographer collects a corpus of data, constructs a formal analysis, and checks the adequacy of that analysis by returning to those data. The inadequacy of this approach derives from the distinction between exploratory and confirmatory inquiry (Giere 1976). The period of exploratory inquiry finds the anthropologist developing working hypotheses and analyses. In this phase of inquiry it is legitimate for the ethnographer to seek recourse to the same corpus of data that was used to derive those working hypotheses and analyses. The period of exploratory inquiry would thus yield testable hypotheses and the beginnings of a theoretical structure. The period of confirmatory inquiry, however, focuses on the testing of well-developed theoretical structures via the hypotheses derived from them. The confirmatory period yields statements about the nature of the phenomena being investigated. Even if no rival theory exists, it is not legitimate to return in the confirmatory period to the same corpus of data used to develop the theory, because this creates a statistical bias toward the confirmation of the hypotheses under test (Kleinbaum and Kupper 1978). When cognitive anthropologists check the adequacy of their analyses using the original data from which the analyses were derived (a formal analysis is "an apparatus for predicting back the data at hand" [Lounsbury 1964:212]), they are doing nothing more than confirming that their working hypotheses are worthy of empirical test. Such a technique does not bear on the validity of the theory underlying those hypotheses.

This appears not to raise any problems for the anthropologist who would return to the field to collect a new sample of the informants' behavior against which to test the analysis. This approach, however, is confounded by the nature of cognitive systems themselves, for cognitive systems are dynamic, continually changing as they alter their fit to an individual's environment (Piaget 1971a; Neisser 1967). The structure and content of an individual's cognized environment is uniquely influenced by experience. Thus between research visits, any (living) informant's system will necessarily undergo some change. The best that can be said, we think, is that present models and research techniques are inadequate for testing the psychological reality of ethnoscientific analyses.

The Sharedness of Cognitive Systems

A major assumption of much cognitive anthropological work is that all members of a society use a shared cognitive system. The explication of cognitive systems of a limited number of informants is therefore seen as sufficient for the explication of the "cultural cognitive system." Although recognition and interest in intracultural variation on the part of cognitive anthropologists has increased in recent years, their treatments of this issue generally assume that all nonsharedness can be handled in terms of cognitive specialization. Werner and Fenton (1973:540–41), for example, acknowledge the problem of intracultural variation and opt to resolve the difficulties it raises for both theory and method by suggesting that ethnoscience models the union, rather than the intersection, of native models. Thus, they suggest that cognitive anthropology models the competence of a "suprainformant." This move fails for three reasons: It glosses over the possibility that the suprainformant's competence might thereby contain contradictory cognitive processes; it incorporates the weaknesses of the ethnoscientific use of the competence versus performance distinction discussed below; and it does not address the question of how specialized cognitive competencies are coordinated in social action.

Not only is a mutually shared cognitive system not a logically necessary condition for the existence of sociocultural systems, but in some instances of social interaction cognitive sharing may actually be undesirable or impossible. Furthermore, empirical evidence concerning the contents of cognitive systems suggests that they are highly idiosyncratic models of the environment. The successful combination of such systems often requires that there be some facilitating mechanism (like ritual; see d'Aquili et al. 1979) at work.

This line of criticism was first developed by Wallace (1961:27–34). His argument is worth a lengthy summary. Wallace argues that a (cognitive) system is defined by two attributes. It is composed of a group of elements which are related such that a change in the state of one produces a nonrandom (or predictable) change in at least one other element in the group. There needs to be at least one sequence that involves all the elements of the group. He considers the question What is the least complex system an ethnographer might describe? And he notes (1961:27):

> Such a system must satisfy the following minimal requirements: first, that two parties A and B, the initiator and the respondent, respectively, interact; second, that each completion of one sequence of interactions be followed, sooner or later, by a repetition of the same sequence. Representing the sets of A by the symbol a_i, those of B by the symbol b_i, and the temporal relationship by the symbol →, to be read "is followed by" we assert that the simplest such system has the following structure:
>
> $$a_i \leftrightarrow b_i$$

Wallace calls this a primary equivalence structure (ES_1). He notes that by definition an ES_1 indicates that A does a_i and then B does b_i, and that when B does b_i, sooner or later A does a_i. He therefore suggests that a secondary equivalence structure (ES_2) would have the following form:

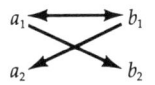

$$\begin{matrix} a_1 & \leftrightarrow & b_1 \\ a_2 & \leftrightarrow & b_2 \end{matrix}$$

Labeling a_1 and b_1 instrumental acts, and a_2 and b_2 consummatory acts, Wallace points out that in an ES_2 the "distinguishing feature is that the consummatory act of each party is released by (but not necessarily exclusively conditional upon) the instrumental act of the other." So in a teleological sense A does a_1 in order to get to do A_2. Likewise for B's doing b_1 and b_2. Hence,

> It is therefore true by definition that neither A nor B will continue to participate in the system unless, first, each perceives that within the limit of the system, his ability to perform his own consumma-

tory act depends on his partner performing his instrumental act; second, that when he performs his own instrumental act, its function is to elicit his partner's instrumental act; third, that he repeatedly performs his own instrumental act. (Wallace 1961:30)

Using only these few assumptions to construct the simplest cognitive maps which satisfy the requirements they set out, Wallace is able to show that there are four "cognitive" maps compatible with the continued existence of a simple system of social interaction, thus demonstrating that it is not logically necessary for the existence of sociocultural systems that their members share a single cognitive map (see Figure 1.1).

What remains is the parallel, though weaker, claim that while it is not logically necessary for members of a sociocultural system to share a cognitive map, such cognitive sharing may be expected to exist. However reasonable this may sound, the bulk of the empirical evidence concerning cognitive systems provides no justification for such a view. As noted before, Piaget (1971a), Neisser (1967), Saltz (1971), and Laughlin and d'Aquili (1974) have all demonstrated that cognitive systems are dynamic entities; they result from a constructive process. The direction that the construction takes is importantly influenced by experience, thus making identity of cognitive systems highly unlikely.

In what sense then, are cognitive systems held in common? Clearly the answer is that they are shared only to the extent that they share structure and process, and some content, and even then sharing requires the use of some processes designed to entrain individual cognitive systems for social action. To attempt to combine the content of cognitive systems in detail requires that the details of at least some of the systems being combined be damaged. As Newell (1973:294, 299) put it in his discussion of studies of information processing:

> Never average over methods. To do so conceals rather than reveals. You get garbage or, even worse,

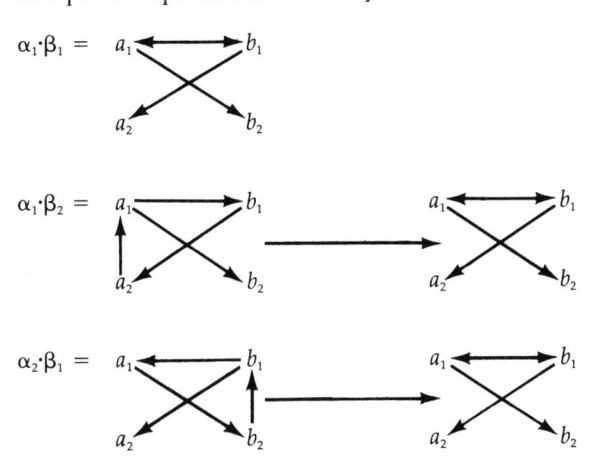

Figure 1.1
*Cognitive Maps for a Simple Social Interaction System**

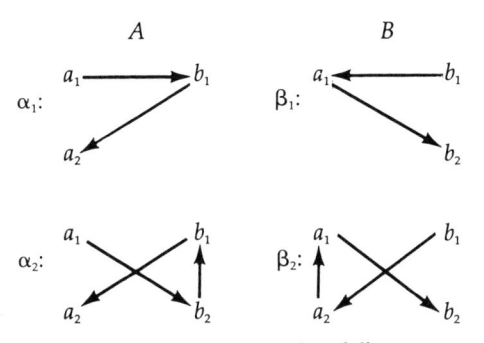

[Diagrams: α_1, α_2 under A; β_1, β_2 under B, showing arrows between a_1, a_2, b_1, b_2]

These maps are to be interpreted as follows:

α_1: A knows that whenever he does a_1, B will respond with b_1, and A will then perform a_2.

α_2: A knows that whenever he does a_1, B will respond with b_2 and then b_1, and A will then perform a_2.

β_1: B knows that whenever he does b_1, A will respond with a_1, and B will then perform b_2.

β_2: B knows that whenever he does b_1, A will respond with a_2 and then a_1, and B will then perform b_2.

Each possible combination of these cognitive maps will yield a structure that is identical with, or logically implies, $^2ES_2$ [a two-person equivalence structure]. Thus:

$\alpha_1 \cdot \beta_1 =$ [diagram]

$\alpha_1 \cdot \beta_2 =$ [diagram] $\longrightarrow$ [diagram]

$\alpha_2 \cdot \beta_1 =$ [diagram] $\longrightarrow$ [diagram]

Figure 1.1 continued
Cognitive Maps for a Simple Social Interaction System*

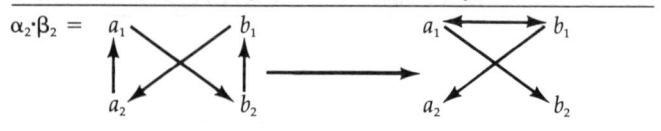

$$\alpha_2 \cdot \beta_2 =$$

From *Culture and Personality* by A. F. C. Wallace. Copyright © 1961. Reprinted with permission of Random House, Inc.

> spurious regularity. . . . The same human subject can adopt many (radically different) methods for the same basic task, depending on goal, background knowledge, and minor details of payoff structure and task texture. . . .

There appears to be no empirical support for the claim that such detailed combination of cognitive systems can be made. There is a growing recognition within ethnoscience itself that cognitive nonsharing may more often be the rule than the exception (Sanday 1968; Werner and Fenton 1973).

The Competence Versus Performance Dichotomy

Much of cognitive anthropology has grown by analogy to linguistic theory. Because of this, cognitive anthropology owes many of its theoretical and methodological concepts to linguistics. Often embedded in the work of cognitive anthropologists are two distinctions borrowed from linguistics. (1) Deep structure versus surface structure and (2) competence versus performance. Frequently, cognitive anthropologists approach their analyses with the suggestion that the behaviors they examine are simply surface manifestations of a deep structural level. Their use of this distinction is relatively nonproblematic. However, the use of the notions of competence and performance in cognitive anthropology incorporates some serious difficulties into the approach.

In his programmatic statements about the study of language, Chomsky (1965, 1972) draws a distinction between a speaker-hearer's competence and his or her performance. By competence Chomsky intends to refer to "the speaker-hearer's knowledge of his language" (Chomsky 1965:4), although Missner (1970) suggests that it refers to the logical features of a linguistic system, and Straight (1976; see also Vandamme 1972) suggests that competence refers to a set of coding principles which linguists abstract from language data. Performance consists of the actual production and comprehension of language by people; "the actual use of language in concrete situations" (Chomsky 1965:4). Cognitive anthropologists have sought to apply this distinction to the study of cognitive systems.

Cognitive anthropology claims to explicate the cognitive calculi used by informants to produce actual behaviors. To maintain that the explication of these models reflects the actual cognitive processes employed by informants is to develop a theory of performance. But to treat the study of "native models" as the explication of some idealized informant's knowledge of his or her culture (*à la* Chomsky), or as the abstraction of the logical features of the environment, or as the development of a set of coding principles for classifying features of the environment, is to develop a theory of competence. From the former it is legitimate to claim that "this is how the informants go about reaching a particular set of observed behaviors." From the latter it is only legitimate to speak of idealized individuals in idealized situations, of coding principles, or of the logical features of the inferred system. But these may have little to do with how behavior is actually produced or how the cognitive system succeeds in modeling the environment. This is so because only when the situation is idealized (and thus vastly simplified) can the processes underlying the production and interpretation of behavior be considered a direct reflection of competence and thus be taken as a unitary set of processes (Straight 1976; Hymes 1974). It appears, in fact, that people necessarily use one set of cognitive processes for interpreting linguistic input and another, distinct set for producing linguistic output. The different neurocognitive processes underlying language production and comprehension in daily life (and by extension the production and interpretation of nonlinguistic behavior) requires that a theory of cognitive performance specify both sets of processes (see Table 1.1).

Unfortunately, cognitive anthropology uses what can only be taken as idealized cognitive models and presents them as systems of cognitive performance. Although it seeks to develop a theory of cognitive performance, the methodological procedures used in cognitive anthropology produce at best accounts of cognitive competence. This confusion is in part responsible for the difficulty cognitive anthropology has in resolving the questions surrounding the issues of cognitive sharing and of psychological reality.

Explanatory Perspective in Cognitive Anthropology

The analysis of social regularities may be pursued on any of a number of levels (psychological, neurological, social, or cultural, for instance) and the power of analyses at different levels may appear to vary widely. However, an adequate and complete explanation of social behavior can be given only by considering all levels of organization present in the interaction of an individual and the environment, both social and physical (see Rubinstein and Laughlin 1977; this position is argued later in this book in Chapters 5, 6, and 8). Insofar as cognitive anthropology attempts to explain social behavior by reference to a single level of analysis, it must be said to lack the explanatory perspective necessary for the full understanding of behavior. This view of the necessity of multilevel explanation in the study of human behavior is critical not only to cognitive anthropology but also to other science (see Wimsatt 1980a, 1980b).

In sum, then, the techniques of cognitive anthropology are useful for anthropological analysis, but these

Table 1.1
*Evidence for a Dichotomy Between Comprehension and Production in Linguistic Theory**

	Comprehension	Production
Ambiguity	Multiple, dynamic	Fortuitous
Attention	More than one input at a time	Only one input at a time
Accessibility of intended meaning	Limited by context and superficial evidence	Unlimited (except via self-monitoring comprehension processes)
Phonological features	Auditory or visual ("taxonomic")	Articulatory (plus auditory and kinesthetic features employed by feedback mechanisms; "systematic")
Flexibility (size of vocabulary, range of sensitivity)	Necessarily great (passive vocabulary superior to active, understand much we do not produce)	Usually rather limited (neither inarticulateness nor inability to mimic hinders comprehension)
Variation	Wide tolerance and high recognizability, large number of styles or registers	Probabalistic or competitive, limited number of styles or registers
Nature of the editing process	Weeding out of implausible interpretations	Backtracking and reformulation on basis of self-monitoring
Neurology	Wernicke's Area (posterior superior temporal)	Broca's Area (posterior inferior frontal)

*After Straight 1976:536.

techniques will not result in a theory of cognitive processes, however much cognitive anthropology may reveal about the formal structure of the perceptual and actional complexes created and manipulated by the mind. The sources and outcomes of human cognition require elaboration in a theory of cognitive processes testable against evidence of the ongoing activity of human perceivers and actors. In the next chapter such a theory is presented as a necessary step in the formulation of a cross-culturally valid way of studying mental models of the environment, which are the psychological underpinnings of sciencing.

Two

Process and Structure in Cognition

This book offers an account of science as cognitive process. The theme throughout is that science involves the deliberate elaboration of biopsychological processes common to science in particular and to cognition in general. As a result, no understanding of science will be complete unless it incorporates, or is at least compatible with, a well-developed understanding of the processes underlying human cognition. We do not argue that we know enough yet about human cognition to arrive at a full and accurate view of science as cognitive process, but we are certain that such a view is essential in order to understand the nature of science.

The Cognized and Operational Environments

Perhaps the major function of the brain is to model reality. This function is carried out so efficiently that we are hardly aware that the world we experience is not objective. Yet there are good reasons to suppose the existence of such an objective world, a world of complex processes in dynamic interaction, changing through time. This is clearest through example. Tiny mesobiota, nematodes, are so numerous that they cover virtually all solid surfaces on earth. Overgaard-Nielsen (see Odum 1971:369) estimated that their population can be so dense that between one and twenty million may live within a square meter of soil. It has been said that if everything on earth but nematodes were to vanish suddenly, we would see the outline of everything through the veneer of nematodes remaining.

With the probable exception of some microbiologists, most people do not conceptualize nematodes in their everyday view of the world. They do not, this is to say, include nematodes as part of their model of reality. Yet we doubt that they would wish to say that because they have not modeled them, nematodes do not exist.

This distinction between the world composed of models within our brains and the world that is objectively real and modelable is fundamental to an attempt to understand how we come to know what we know. For this reason we make a distinction between the *cognized environment* and the *operational environment*.

The cognized environment (E_c) consists of all the information modeled in an individual's nervous system through the operation of which the individual recognizes, processes information about, and responds to the operational environment (E_o). This distinction is pivotal, and it is helpful to explore briefly some of its more interesting implications.

The cognized environment develops; it grows, in a biological sense, as a specific function of the general processes of individual development (Piaget 1971a, 1977). In the process of acquiring and developing structure, neural models are constructed within an individual's nervous system. As a result the cognitive principles directing the construction of neural models place constraints on the completeness and accuracy of the individual's cognitive representation of the operational environment.

Included in the potentially modelable operational environment are many objects and events integral and internal to the individual. There are many aspects of our own organism that we commonly do not model, for example, the microbiota that inhabit our alimentary canal and aid our digestion, the melanocytes that produce our skin and hair color, the billions of molecules that compose our bodies, or even most of our internal organs. Thus, we can distinguish among (1) the operational environment specific or internal to an individual, (2) the operational environment external to an individual, and (3) the cognized self (a special aspect of the cognized environment), which is an attempt to model various self-relevant portions of an individual's internal and external operational environments.

Just as there are aspects of an individual's internal and external operational environments that are not modeled in the cognized environment, an individual may construct models that have little or no correspondence to the operational environment. Examples of such models with varying degrees of correspondence to the operational environment can be taken from a wide variety of domains. Some entities, like gods, are believed present in the operational environment by a majority of people, and some fraction of those claim to have experienced them. Other entities, like leprechauns and nymphs, are thought by some people to be real because others claim to have seen them. And, some entities, like hobbits or the Easter Bunny, are simply fictitious. A number of processes the operation of which cannot now be confirmed, like transmigration of souls or transubstantiation, also can be included as examples of such models. Some models may combine veridicality and delusion in adaptation to the operational environment (see Laughlin and Stephens 1980).

To some degree the structure and content of an individual's cognized environment models the elements and relations among elements in the operational environment. As an individual's cognized environment develops, it is at different times more or less isomorphic with the operational environment. This notion of isomorphism is important, and it has a special sense in our discussion.

Isomorphism refers to the correspondence between the elements and relations constituting a particular system, and the elements and relations constituting another system of different form. Of course, while there exists a correspondence between the organization of one system vis-à-vis another, the actual elements and relations constituting each system are different. When we describe the relationship between nematodes and bacteria, the elements and relations constituting one system (our descriptions) are signs (words, numbers, and the like) and logical relations, while the elements and relations constituting the other system are organisms and biotic relations. Yet the way we arrange the signs and logical relations in the former system in some sense matches—is isomorphic with—the arrangements of organisms and biotic relations in the world. Thus, for example, while the brain models the operational environment by discrete arrangement of neural tracts, the aspects of the operational environment being modeled need have nothing to do with these neural tracts. Nonetheless, there exists a kind of correspondence between the arrangement of neural tracts composing the cognized environment and the arrangement of corresponding aspects of the operational environment (Pribram 1971).

This relationship may be expressed as follows. A relationship between the two systems S and S' is isomorphic just in case there is the same number of elements in the two systems and there is a one-to-one correspondence of the relations among these elements in the two systems.

Complete isomorphism means that the organization of one system completely maps the organization of another, and vice versa. But complete isomorphism rarely exists. Thus, the cognized environment is never more than partially isomorphic with the operational environment. To illustrate, the relationship between signs and symbols in a road map, on the one hand, and the towns, roads, and other objects in the geographic area covered by the map, on the other hand, is partial at best. The road map is then only partially isomorphic with the operational environment being modeled. *Our concern, then, is with the partial mapping of one system by another. By extension, the entire cognized environment is at best only partially isomorphic with the operational environment.*

This still does not fully describe the complexity of

the situation. Because the construction of knowledge about the operational environment in the form of cognized environment models is a biological phenomenon (the nervous system provides this function for an individual), it is open to selection. Hence, the cognized environment is adaptively isomorphic with the operational environment when the degree of "fit" between the cognized environment model of the operational environment leads to the individual's survival and production of offspring.

Changes in the operational environment may be of three general types. They may be caused by unique events, they may be ongoing, and more or less cumulative (such as the increase in population density or the normal acquisition of language by a child), or they may occur in a cycle that has a more-or-less predictable range of frequency and amplitude (such as periods of drought which occur annually, or typhoons and hurricanes). The maintenance of adaptive isomorphism between an individual's cognized environment and a changing operational environment of the first two types may be accomplished by simple accommodation of the cognized environment to the operational environment facts. In the face of cyclic shifts in the operational environment, however, there is the possibility of maintaining adaptive cognitive isomorphism by modeling the cyclic shifts themselves. This tendency had been termed *diaphasis* by Laughlin and Brady (1978:21). One particularly intriguing example of diaphasis in relation to science discussed later is that of the fluctuation in level of cognitive complexity in response to fluctuations in operational environment novelty.

The cognized environment can model itself; it may construct models of its own models. Just as the cognized environment models are typically partially isomorphic with the aspects of the operational environment they model, the secondary model will be only partially isomorphic with the primary model, and so on. With each remove from the initial modeling, the more abstract model loses some closeness of fit with the system being modeled. This does not necessarily mean that abstract models are less adaptive. In fact, the opposite may be the case. It was once thought that cognition was based upon language (Watson 1920). The present understanding is that cognition is evolutionarily and ontogenetically prior to language, and that language models cognition. Nonetheless, language serves as an important process for bringing about the adaptive isomorphism of cognized models with the operational environment. This is despite the fact that the modeling of cognition by linguistic expression results in the loss of some information. Cognition is a simplified representation of the operational environment, and language is a simplification of this simplification (see Table 2.1).

The nervous system operates on the operational environment as a unit, but many of the information-processing systems operating within the nervous system do so outside of consciousness. We may not be aware of the structure or of the information being processed by that structure, or we may be aware of the information but not aware of the mediating structure. Typically the conscious aspect of the cognized environment constructs

Table 2.1
Some Relations of Isomorphism in E_c/E_o Processes

(1)	Complete	Total mapping of E_o by E_c
(2)	Partial	Incomplete mapping of E_o by E_c
(3)	Adaptive	E_c/E_o "fit" sufficient to lead to production of offspring
(4)	Developmental	Accuracy of E_c mapping of E_o continually adjusted by equilibration
(5)	Diaphatic	E_c models repetitive shifts in E_o by recurrent shifts in its structure and content
(6)	Self	E_c models of an individual's own organism as an element in the E_c (the "cognized self")
(7)	Secondary	The fit between E_c models of E_c models

models of the individual's own nervous system function and structure. This cognized environment's modeling of itself produces the "ego." In modeling its own functioning, the cognized environment inevitably resorts to transposition. The ego is a simplification of self, the notion of "scientific method" a simplification of processes of everyday cognition, and so on. When an individual operates using a cognized environmental model, *the model may constrain the functioning of the system to those simplified elements and relations*. For this reason both our knowledge of the operational environment and our knowledge of the operating structures are fallible.

The cognized environment operates as a system (Harvey et al. 1961; Piaget 1971a, Laughlin and d'Aquili 1974). When confronted by anomalous input from the operational environment, that system, or any subsystem thereof, functions for as long as possible in ways that maintain its integrity. It does this in a variety of ways, including distorting information for assimilation into the system (Wallace 1957; Turiel 1966) rejecting the information (Hastorf and Cantril 1954; Festinger et al. 1956), and holding the information within memory without assimilating it into the cognized environment.

The cognized environment may remain relatively open to information input and action output vis-à-vis the operational environment. This occurs through a feedback-feedforward system involving sensory input and motor systems in complex tests of the goodness of fit between the cognized environment models and the operational environment (Laughlin and d'Aquili 1974; Miller et al. 1960). This testing, evaluating, and modifying of cognitive models can maintain their adaptive isomorphism. The complex process that serves to maintain the internal integrity of the cognized environment while simultaneously allowing the system to undergo change is equilibration (Piaget 1977).

Activity in Cognized Environment Modeling

Equilibrating transformations can be divided into three types. A *developmental transformation* increases or decreases the complexity of the operating structures. Within any level of cognitive complexity, information may be processed by a partial set of neural subsystems entrained in a functional array (Luria 1976). A change in the organization of an array of neural subsystems is a *surface transformation*. Such surface transformations are characterized by a system remaining at a single level of organizational complexity, yet undergoing a repatterning of its arrangement of elements, or a shift in the value placed upon specific elements. A very negative view, for example, may be transformed into a positive view yet have the same structural complexity.

Changes in the information passing through the system are *sensory transformations*. To a significant degree, it is the structure which determines the nature of the information which enters itself. Thus, the structural characteristics and developmental level of the individual will define in part the information he or she "sees" and the mode through which that information is integrated into the system. There appears to be a general tendency for the system to distort the information to fit that already encoded and to process it in terms of the system's complexity of organization.

The disparity between individuals' abilities to experience events and their relative inabilities to directly experience cause-effect relations is the "zone of uncertainty" of the cognized environment. This follows from the inevitable partial isomorphism of cognized models with the operational environment. The ancient Egyptians, for example, were aware of the yearly flooding of the Nile Valley but had no direct knowledge of the cause of that flooding. Knowing that we inevitably die does not explain why we die, nor does it entail an experience of the agent of our death. Generally, however, if a salient event has no directly perceptible cause, one will be provided by the conceptual system. This is due to an inherent and unavoidable organization of events into cause-effect sequences, even if some of the "events" are hypothetical or causally unrelated (Hume 1739, 1748;

d'Aquili 1972). The ancient Egyptians attributed the annual flooding to gods; many cultures attribute death to sorcery, possession, and the like. Once formulated, such conceptual systems will operate to maintain their integrity, at times in the face of overwhelming disconfirmation. The cognized environment serves as a way of interpreting inputs which are themselves already abstractions constructed from events in the operational environment.

Because the cognized environment is a dynamic system, it encompasses and models at least four levels of organization: neurological, cognitive, societal infrastructure, and surface structure (see Figure 2.1). The

Figure 2.1
*Levels of Structure Operating in the Human Organism as an Actor in a Social System**

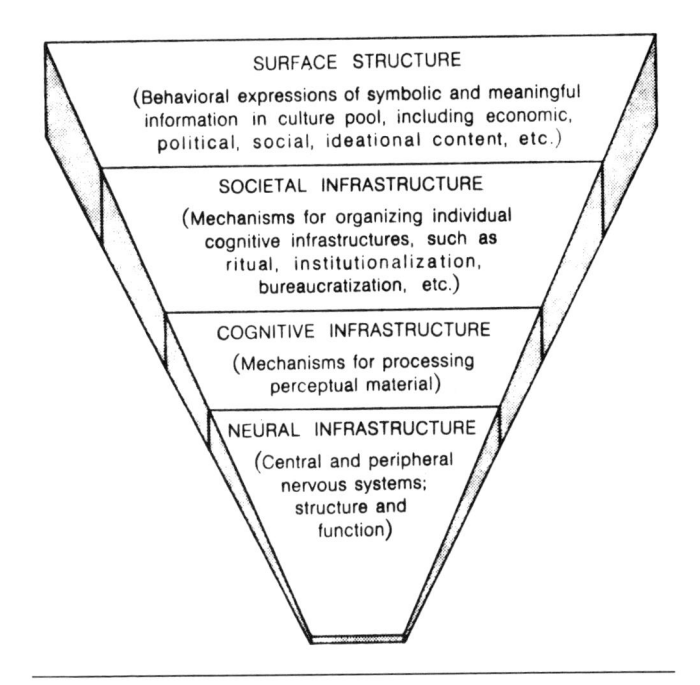

SURFACE STRUCTURE
(Behavioral expressions of symbolic and meaningful information in culture pool, including economic, political, social, ideational content, etc.)

SOCIETAL INFRASTRUCTURE
(Mechanisms for organizing individual cognitive infrastructures, such as ritual, institutionalization, bureaucratization, etc.)

COGNITIVE INFRASTRUCTURE
(Mechanisms for processing perceptual material)

NEURAL INFRASTRUCTURE
(Central and peripheral nervous systems; structure and function)

*After Laughlin and Brady 1978:4.

cognized environment affects and is affected by the adjoining levels. It organizes the neurological structures into functional systems and it is composed of those neurological structures, and it adapts to the operational environment structures external to it (Rubinstein and Laughlin 1977; Laughlin and Brady 1978).

Each level of structural organization operates as a buffer and a regulator between structures internal to itself and structures external to it. Thus the cognized environment operates as a regulator between the operational environment, and the neural structures and systems from which the cognized environment is constituted. This has several important implications for the study of science:

1. The integrity of any structure is directly related to how permeable it is. Total permeability is equivalent to nonstructure.

2. The neurological structures constituting the cognized environment may be affected directly by the operational environment, without mediation by the cognized environment.

3. The concept of "cognitive boundary" refers to the resistance of the cognized environment (or some subsystem in it) to such direct influence.

4. Indirect influence on the cognized environment may be through information or changes in structural organization. Depending upon its nature, such indirect influence may result in a decrease in cognitive complexity, or information-processing capacity, or alternatively in the simple assimilation of changes into the cognized environment models.

5. Adaptive cognized environments may under some conditions be both directly and indirectly affected by changes in the operational environment. A totally nonpermeable cognized environment will not long facilitate survival in a changing environment, while one too open to the influence of the environment is vulnerable to disruption and dissolution. The notion of diaphasis, discussed earlier, implies that an adaptive cognized environment will be appropriately flexible in a variety of contexts.

6. All else being equal, the more complex the organization of a cognized environment, the less permeable it is to direct structural influence. More complex cognized environments will admit, assimilate, organize, and even hold separate more information from the operational environment than simpler cognized environments will. Thus, complexly organized cognized environments are more stable and flexible and are less vulnerable to structural change from the direct influence of the operational environment.

7. Cognized environments are sensitive to environmental influence in different degrees according to the subsystems considered. Each subsystem level has stimulus-defining properties. Principal influences, however, upon the content and functioning of the component systems are exerted by the neurocognitive and external operational environments.

8. A minimal amount of operational environment complexity is necessary to stimulate the optimal functioning of each subsystem of the cognized environment (Rubinstein and Tax, 1981). However, an overabundance of environmental complexity may cause a decrement in the functional capacity of the system (Schroder et al. 1967). The vulnerability of each subsystem, and of the cognized environment as a whole, is contingent upon its complexity and upon the amount of environmental complexity. The relation between these two levels of organization takes the form of mutually interacting U-curve functions (see Figure 2.2). When a strong influence is exerted on the cognized environment (either directly or indirectly) by large amounts of environmental complexity, it closes down and ceases to deal with complex or contradictory input, its functioning focuses on the concrete, and, perhaps, it shows an unusually high rate of internal activity.

Operational and Cognized Logics

Abraham Kaplan (1964) makes a useful distinction between logic-in-use and reconstructed logic:

The word "logic" is one of those words like "physiology" and "history," which is used both for a discipline and its subject matter. We all have physiologies and histories, and some of us also think and write about these things. Similarly, scientists and philosophers use a logic—they have a cognitive style which is more or less logical—and some of them also formulate it explicitly. I call the former the *logic-in-use*, and the latter the *reconstructed logic*. (Kaplan 1964:8)

Kaplan's logic-in-use refers to operations actually performed by the scientist's or philosopher's cognitive system upon information. For consistency with our earlier discussion, we call this operational logic. A recon-

Figure 2.2
General Relationship Between Environmental and Cognitive Complexity*

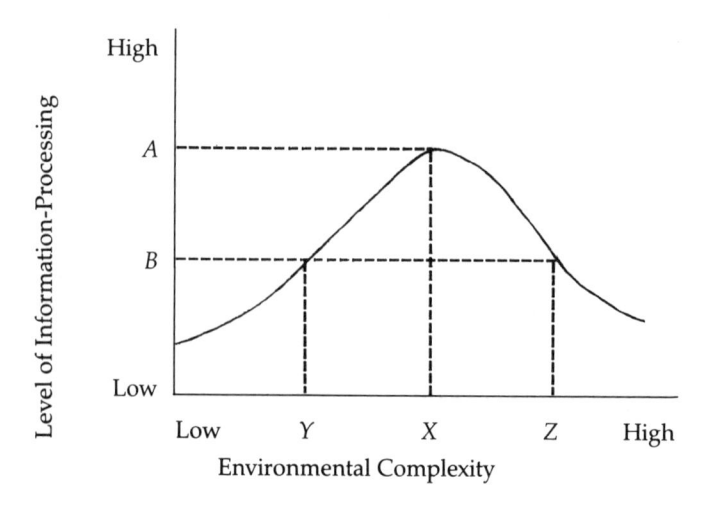

**From HUMAN INFORMATION PROCESSING by Harold M. Schroder, et al. Copyright © 1967 by Holt, Rinehart, and Winston, Inc. Reprinted by permission of Holt, Rinehart and Winston, CBS College Publishing.

structed logic is a person's cognized model of the information-processing occuring within his or her cognized environment; we call this cognized logic.

Extrapolating from the general model, several interesting things may be added to the notions of cognized and operational logics:

1. Just as the ego is the product of the cognized environment's modeling of its own processes, so a cognized logic is the product of the cognized environment's modeling of the principles of its own information-processing. For any domain there is only one operational logic, but potentially many cognized logics.

2. If, however, an individual is operating upon information via conscious application of a cognized secondary model of that process, the model will likely constrain the operation of the process modeled. For example, an individual consciously applying a cognized logic will (often severely) constrain processing to the parameters cognized (cf. Wimsatt 1980b). By shifting cognized logics, the individual may shift the constraints upon the operational system (Tart 1975).

3. Some cognized logics are more fully isomorphic with the operational structures and processes that they model than are others. We take it that cognized logics grounded in an empirical examination of operational logics tend to be more fully isomorphic with them than nonempirically (and often normatively) derived cognized logics. Cognized logics based upon the examination of the neurocognitive systems underlying information-processing will be more accurate than those based solely upon *a priori* programmatic or normative criteria.

4. The development of the fields of mathematics and philosophy of logic has generally reflected a progressive development of the cognized environment's operational logic (Beth and Piaget 1966). This development exhibits the trend toward increased awareness on the part of Homo sapiens of the dynamics of their own organism. Yet because all knowledge is fallible, particularly conscious knowledge, it is clear that there is much left to be learned

about human information-processing. This will undoubtedly result in the increased complexity of cognized logics.

5. A cognized logic may facilitate or impede inquiry, depending upon its normative status and upon its relation to the operational logic of reference. Thus, a normative cognized logic may define a particular inquiry as "invalid," "inappropriate," "unethical," and so on. Thereby it may impede inquiry into a particular domain (see Rubinstein 1984, Welsch 1983, and Hufford 1982, 1984). On the other hand, such a logic might effectively inhibit premature closure of the theory and thereby facilitate accommodation of previously anomalous data that otherwise might have been ignored or rejected within the theory. Appeal to a normative cognized logic is a common criticism in scientific discussion: "Proper procedures were not followed," "There was no control group," "This problem is uninteresting," and so on.

6. Cognized logics participate in the definition of scientific paradigms. Because of this, such logics tend to become reified, much like the systems of norms characterizing societies in general. As such, paradigmatic cognized logics may ultimately stifle creative research (see, e.g., Toulmin and Goodfield 1961: chap. 6). There seem to be two checks that can be placed upon this process. First, scientists can be prepared to "see" beyond the constraints imposed by any cognized logic. Second, the most complex and fully isomorphic model of the operational environment possible for use as a cognized logic must be developed.

Thomas Blackburn (1971:1007) has spoken to the first of these alternatives:

> What is urgently needed is a science that can comprehend complex systems without, or with a minimum of, abstractions. To "see" a complex system as an organic whole requires an act of trained intuition, just as seeing order in a welter of numerical data does. . . . The intuitive knowledge essential to

a full understanding of complex systems can be encouraged and prepared for by: (i) training scientists to be aware of sensuous clues about their surroundings; (ii) insisting upon sensuous knowledge as part of the intellectual structure of science; and (iii) approaching complex systems openly, respecting their organic complexity before choosing an abstract quantification space into which to project them.

We take "trained intuition" to mean that analysts are capable of transcending constraints imposed on their own cognized environments by normatively accepted cognized logics, and hence entering into direct and complex interaction with the operational environment, constrained only by the limits of their operational structures. This is very much like "seeing" in the Buddhist sense, viewing an aspect of the operational environment without projection of normative knowledge associated with that aspect. The result of such activity ought to be a more effective and complex modeling of the operational environment, and thus the development of more fully isomorphic models of that environment. Certainly one way of facilitating this inquiry is by applying a cognized logic that recognizes the limits it places on inquiry. It is to that end that the remainder of this book is directed.

Three

Multivariable Cognitive Development

The preceding chapters raised some basic issues that constrain an empirically grounded philosophy of science, and initially defined some of the central concepts of this study. The central goal of science is to bring cognized models of the operational environment into the most adaptive possible alignment with actual entities and relations within the operational environment. This process of approximation is the natural by-product of human cognition. As such, it is constrained by principles of neurocognitive information-processing and storage.

This chapter explores more fully the principles of cognition upon which subsequent discussion is based. It examines the development of cognitive systems from the points of view of Jean Piaget's genetic epistemology and neo-Piagetian developmental theory. The importance of these theories to social science is sketched after brief outlines of each.

The Concept of Structure

The concept of structure is at once one of the most important and most difficult notions to communicate. Cole and Scribner (1975:253–58) note that cross-cultural studies of cognition vary along several dimensions. One is the dimension of "content" versus "process." This turns out to be a major distinction between anthropological and psychological studies in "cross-cultural psychology." The distinction rests on whether the approach focuses on the information contained in a cognitive system or on how that information is organized and manipulated. The content approach is exemplified in the work of anthropologists (e.g., Tyler 1969; Spradley 1970, 1972) who focus mainly on what is processed. Psychological work in the cross-cultural study of human cognition has focused mainly on process, the manner in which information is used by individuals (see Price-Williams 1969; Kagen et al. 1973).

The distinction between content and process approaches to the study of human cognition yields a number of other dichotomies. On the one hand, content theorists concentrate on explicating the great array of differences in human cognition that exist among cultures. This emphasis results from the fact that they focus on different cultural materials as their examinations move from one culture to the next. The picture that results from work done by the content school is one of extreme diversity in human cognition. On the other hand, process theorists provide a picture of great similarity in human cognition from one culture to the next. This is due, in part, to the fact that they focus on materials that can be found in human cognition throughout all societies. The apparent difficulty in forming the results of cross-cultural research in human cognition into a coherent whole, noted by Paredes and Hepburn (1976; they called it the "culture and cognition paradox"), resides in the differing assumptions taken by the process and content approaches to the study of human cognition.

Structure in human cognition is the organization of thought. It should be distinguished from content, or that which is thought about. Structure is the information-processing system and content is the information processed.

Piaget (1970:5) defines structure as a system of transformations. Such systems involve three components: wholeness, transformation, and self-regulation. Wholeness refers to the fact that the elements of thought are subordinate to its principles of composition. These principles of composition, or transformations, are the actual or potential mechanisms of organization operating within the cognitive system. Self-regulation labels the fact that cognitive structures are active neurobiological processes, which maintain specifiable boundaries, interact with an operational environment, and are capable of change within their biological and formal constraints.

A cognitive structure, then, is the formal organization of an aspect of the cognized environment. It is simultaneously capable of maintaining a consistent identity while changing through interaction with the environment. In Piagetian terms, cognitive structure is virtually synonymous with adaptive intelligence (Feldman et al. 1974). The structure defines and selects salient incoming information and provides the organizational substrate for behavior.

Cognitive structures vary in their degree of complexity, stability, and adaptability. These dimensions are usually seen as describing continua from concrete to abstract. Position on any continuum is determined by the complexity of the internal organization of the cognitive structure. More complex structures are more stable and hence more adaptive than are more simple structures (Langer 1969). The definition of and responses to "objective" reality are determined by the structure of the apperceiving cognitive system. It follows that complex cognitive systems can model the operational environment in more complex terms than simple cognitive structures do, and they may adapt to the environment in more complex and flexible ways. Differences in structure at the level of individual cognized environments have consequences not only for the behavior of individuals but also for any combination of individuals, including social institutions.

Biological Bases of Structure

Cognitive structures grow out of, exist in, and are continuous with the biology of living organisms (Count 1973, n.d.; Fishbein 1976). Piaget (1971a) argues that cognition is coextensive with the "functional invariants" of any living organism: (1) organization of its own internal systems, and (2) adaptation to the surrounding environment.

Some aspects of cognition have a genetic base, and the principal structures of thought are constructed from this base through interaction with the operational environment (Piaget 1971a; Seligman and Hager 1972; Pribram 1971). This is carried out through the organic functioning of neurocognitive systems in relation to the operational environment. This process allows the continual reappraisal and restructuring of the internal organization of an individual. Cognitive structures at all levels of complexity are biological phenomena continuous with other biological functions, but differing in terms of complexity, scope, and adaptive significance.

The Concept of Development

Understanding cognitive structure is a task that requires a historical perspective. While atemporal questions may be usefully asked, the resulting synchronic data are arbitrary points abstracted from ongoing process (Bateson 1979). Static structures are either the artifacts of cross-sectional analysis, or instances of structures whose development has been arrested or completed at a given level of organization. Development is distinguishable from related concepts such as growth or change because, within a range, it has a fixed direction and connotes a reorganization rather than the simple addition or replacement of elements (Waddington 1957). Development proceeds from states that are relatively undifferentiated, global, and rigid to states that are relatively more differentiated, integrated, abstract, and flexible (Werner 1957; Schroder et al. 1967).

Cognitive structures also demonstrate certain stage characteristics (Flavell 1963). This means that at some points in time they form a coherent, complete structure which is more complex than preceding structures and which incorporates these preceding structures within its own organization. Thus, the continuity of identity is maintained in development while rules of organization are elaborated and extended. The fundamental mechanism of development is equilibration (Piaget 1977), the dynamic balance created when the individual assimilates sensory input into its existing organization while accommodating that organization to the characteristics of the input. When in balance, the assimilation-accommodation interplay allows reorganization of the internal structure of the cognized environment, thus making its modeling of the operational environment richer and more complex.

Structural Developmental Variables

This discussion of structural developmental variables is restricted to three theories of cognitive development which are closely related to each other and which have been explored in a rich literature. These approaches each conform to the use of structure described above, but each is concerned with the organization of thought in a separate content area. These are: (1) the genetic epistemology of Jean Piaget, (2) the conceptual systems theory of Harvey, Hunt, and Schroder, and (3) the moral judgment theory of Lawrence Kohlberg. Our overview of these theories is necessarily brief, referencing other extended reviews where more complete treatments can be found.

Piaget's Genetic Epistemology

The principal body of theory and research relevant to this stage of our discussion is Piaget's genetic epistemology. Concerned with the ontogenesis of the structure of knowing, Piaget's approach is primarily addressed to the epistemology of physical reality. His orientation assumes an intrinsically active organism, functionally continuous with nonhuman animals. Piaget and his collaborators have produced many empirical and theoretical studies on subjects such as the development of knowledge about number, geometry, space, time, perception, mental imagery, memory, and causality (Flavell 1963; Ginsburg and Opper 1969), as well as a substantial literature addressing the general mechanisms of development (Piaget 1971b, 1977).

Piaget identifies four major periods in the development of intelligence: (1) sensory motor, (2) preoperational, (3) concrete operational, and (4) formal operational thought. Operations are reversible actions that can be internalized. Beginning with the sensory motor period (which itself has six substages), intelligence consists of coordinations between the sensory and motor systems. All thought for about the first year and a half of life takes place within this system of coordinations. The coordinations accomplished during the sensory motor period are reconstructed at the level of representational intelligence during this period, and during the following, concrete operational, period. Appearing around the age of seven or eight among Euro-American children, concrete operations signal the arrival of what is typically considered adult thought.

Before concrete operational thought is reached, an individual operates on the operational environment in terms of its figural aspects and focuses particularly on one aspect of the stimulus situation at a time. Coordinations between various aspects of a stimulus are lacking, limiting the individual's adaptive flexibility. One indication of this level is failure to understand concepts that require such coordinations. Concrete operations involve the construction of this capacity and constitute the appearance of logical intelligence. The final period, formal operations, is characterized by the capacity for reasoning from propositions and constitutes genuinely abstract intelligence (Piaget 1972; Piaget and Inhelder 1969).

The rate of development through these stages is variable, but its sequence is fixed. Not all or even the majority of individuals in any society need reach formal operations (Piaget 1972; Dasen 1972, 1977). Development may be uneven, taking place in very specific areas of experience and not in others. As a descriptive model of development, the Piagetian view has received wide support from research in both Western and non-Western societies.

Central to this view of the development of cognitive stages are the dual functional invariants of (1) organization and (2) equilibration. These have been treated at length by Piaget (1971a, 1970) and by reviewers of his work (Flavell 1963; Furth 1969; Langer 1969; Phillips 1969; Radford and Burton 1974), and they are simply outlined here.

Organization

Piaget's attention to the active nature of cognition is expressed in his central concern with its structure and with the ramifications of cognitive structuring for environment—organism interaction (Radford and Burton 1974:179-82). At all levels, cognition displays structuring, which is the result of the functional invariant of organization—the tendency on the part of all living organisms to form their systems and subsystems into hierarchically arranged structures. Organization provides the infrastructure that guides an individual's interaction with the operational environment. It is important that both the active nature of cognition and the infrastructure itself are necessarily present throughout development. The essential point is that at all stages of development the individual acts on the environment based on the infrastructural organization. A mature individual acts no more on the environment than an infant does. What changes through development is the range and scope of the actions, and hence the effects of those actions.

As Furth (1969:263) notes, organization is the most general expression of form in biological organisms. It implies a systemic whole, a totality in which elements are related to one another and to the larger whole as well. Living organizations tend to conserve their own structure, while at the same time extending the range of the environment to which that structure applies. This dual process of conservation of structure and structural expansion is partly the result of organization and partly the result of equilibration.

Equilibration

Equilibration is responsible for establishing the dialectic that underlies structural elaboration within the organism in the same way that organization underlies the creation and imposition of structure on an organism's systems and activities. The basic tension created by equilibration (sometimes called adaptation) results from the complementary processes of assimilation and accommodation.

Assimilation occurs when an organism uses and incorporates into its structure something from its environment. Assimilation is the

> incorporating process of an operative action. A taking in of environmental data, not in a causal, mechanistic sense, but as a function of an internal structure that by its own nature seeks activity through assimilation of potential material from the environment. (Furth 1969:260)

That is, assimilation is the process by which aspects of the environment are acted upon in such a way that they become incorporated into an organism's structure.

A simple biological example of assimilation is the digestion of food. During digestion an organism acts on a piece of its physical environment and then incorporates it into its own structure. In cognition every encounter with environmental objects necessarily involves some kind of cognitive structuring of the perception of encountered objects so that they are understood in a fashion consistent with the existing cognized model of the environment.

Interaction between an organism and its environment during which the structure of the organism is changed to fit the environment is accommodation.

> Accommodation applies a general structure to a particular situation; as such, it always contains some

element of newness. In a restricted sense, accommodation to a new situation leads to the differentiation of a previous structure and thus the emergence of new structures. (Furth 1969:259)

Accommodation requires the alteration of existing structure in a way that creates a better "fit" between the organism's structure and the environment. For example, after eating, an organism must adjust to the peculiar aspects of the ingested food in order to digest it. So after a large meal one's stomach needs to alter its structural proportions (thus becoming distended) in a way different from that required for a smaller meal. On the cognitive level, variation in an organism's environment is accommodated by construction of new cognitive structures (or substructures), thus facilitating the use of novel information. (See Chapter 2 for further discussion of these activities in human cognition.)

A number of people have stressed that for Piaget intelligence is adaptation and that the development of human intelligence is characterized by shifts from one fairly stable pattern of cognitive structuring to another (e.g., Feldman et al. 1974). When, however, either assimilation or accommodation accounts for an unusually high proportion of the activity of a cognitive system, development may be arrested at a particular level of organization. Alone, assimilation distorts sensory input to fit the expectations of the individual, while by itself accommodation prescriptively molds the individual's perceptions and behaviors to match the perceived environmental demands. Development requires both processes. Structures modify themselves through interaction as a function of their existing state and the nature of the surrounding environment.

Conceptual Systems Theory

A second approach of interest is the "conceptual systems theory" developed by Harvey, Hunt, and Schroder (1961; Schroder et al. 1967; Schroder and Suefeld 1971). Conceptual systems theory considers cognition in terms of information-processing. A stage developmental theory, it is concerned with the progressive organization of thought for social interaction. Stages are viewed as arising out of and incorporating previous stages in an invariant sequence. Each stage is differentiated from the preceding stage by greater structural complexity. Development proceeds from an initial undifferentiated stage to a stage that is highly articulated and organized. The higher stages are more efficient and comprehensive information-processing structures and can deal with more diverse information. These higher stages are considered also to be more stable and generally more adaptive than the stages that precede them.

In conceptual systems theory, structural complexity is determined by two variables: dimensions and rules. Dimensions refers to the number of units or parts of information considered by an individual during information-processing. Thus it denotes the ways in which a set of stimuli can be ordered or scaled. Dimensions are, then, ordering principles or categories used by individuals to interpret environmental input. Each information-processing structure (or conceptual system) is in part made up of dimensions representing independent attributes along which stimuli can be ordered. Rules are of two types: (1) mixed rules, which are rigid, minimally modifiable guides for information-processing; and (2) emergent rules, highly flexible guides that are capable of generating many and new perspectives.

Structural complexity is a measure of how well integrated an information-processing structure is. It depends upon both the dimensions and the rules it includes. Yet,

> [the] number of dimensions is not necessarily related to the integrative complexity of the conceptual structures, but the greater the number of dimensions, the more likely is the development of integratively complex connection rules. Low integration index is roughly synonymous with a hierarchical form of integration, in which rules or programs are

fixed.... High integration index structures have more connections between rules; that is, they have more schemata for forming new hierarchies, which are generated as alternate perceptions of further rules for comparing outcomes. (Schroder et al. 1967:7)

Conceptual structures of differing complexity have different modes of information-processing. Schroder (1971; Schroder et al. 1967) distinguishes four stages of structural development, and five structural patterns. These stages and patterns, which correspond roughly to the stages described by Piaget and his associates, distinguish between information-processing employing lower levels of structure (which tend to be concrete, to orient toward external standards, to use categorical thinking, and to avoid ambiguity and conflict) and information-processing employing higher levels of structuring (which are flexible and can generate multiple perspectives on and solutions to particular problems [see Figures 3.1 and 3.2]).

Schroder (1971:257) says of the lower levels of structuring that they show a

> tendency to standardize judgements in a novel situation; a greater inability to interrelate perspectives; a poorer delineation between means and ends; the availabilies of fewer pathways for achieving ends; a poorer capacity to act "as if" and to understand the other's perspectives; and less potential to perceive the self as causal agent in interaction with the environment.

The work of the conceptual systems group has interesting ramifications for social science. This group has demonstrated that cognitive systems function differently at different levels of organizational complexity and that this functional level is sensitive to the complexity and quantity of information input from the environment. There is an optimal level of input from the environment appropriate for each structural level. (For further discussion of this relationship, see Schroder et al. 1967, and the discussion in Chapter 2, "Activity in Cognized Environment Modeling.") Too little stimulation fails to activate the system to optimal levels of functioning, and too much stimulation similarly promotes a deficit in the flexible processing of information. Each progressive level of organization demonstrates a capacity for processing and integrating greater amounts of information and a

Figure 3.1
Lower Levels of Conceptual Structuring*

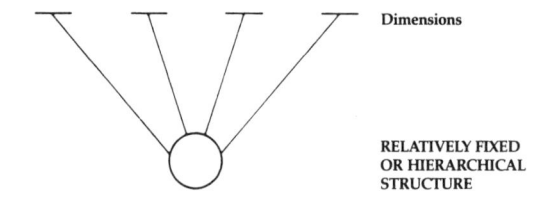

Dimensions

RELATIVELY FIXED OR HIERARCHICAL STRUCTURE

A. *Multidimensional, Single-rule Structure*

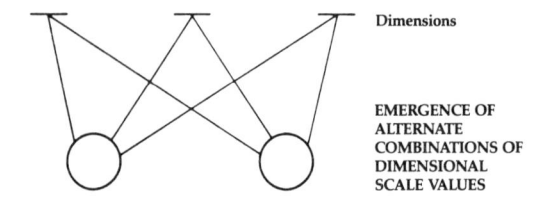

Dimensions

EMERGENCE OF ALTERNATE COMBINATIONS OF DIMENSIONAL SCALE VALUES

B. *Multidimensional, Multirule Structure*

Multivariable Cognitive Development   **49**

Figure 3.2
*Higher Levels of Conceptual Structuring**

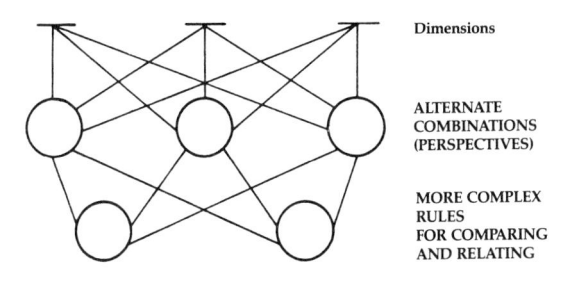

Dimensions

**ALTERNATE
COMBINATIONS
(PERSPECTIVES)**

**MORE COMPLEX
RULES
FOR COMPARING
AND RELATING**

A. Multidimensional, Multiconnected Rule Structure of
Moderately High Integration

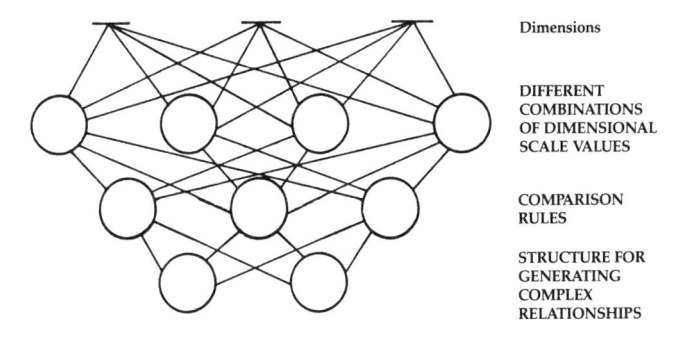

Dimensions

**DIFFERENT
COMBINATIONS
OF DIMENSIONAL
SCALE VALUES**

**COMPARISON
RULES**

**STRUCTURE FOR
GENERATING
COMPLEX
RELATIONSHIPS**

B. Multidimensional, Multiconnected Rule Structure of High Integration

Figures 3.1 and 3.2 are from *Human Information Processing* by Harold M. Schroder, et al. Copyright © 1967 by Holt, Rinehart and Winston, CBS College Publishing.

greater resistance to stress. It is important that when stress (or environmental complexity) becomes so great that the structures of the system begin to collapse, they do so in an orderly fashion and in a way that preserves for as long as possible as much of the integrity of the structure as possible. This phenomenon of "structural collapse" (Rubinstein 1979a) has clear parallels for processes of change in science, discussed in the next chapter. Also of interest is the fact that a small group's adaptive capacity is limited to the level of its most individually complex members (Laughlin and Brady 1978).

Kohlberg's Theory of Moral Development

The final system discussed here was developed by Lawrence Kohlberg. Kohlberg's approach is derived from Piaget's work on the development of judgments concerning moral and ethical questions (Kohlberg 1963, 1969). This approach has been criticized on a number of grounds, particularly its potential cultural bias (Gilligan 1982, Simpson 1974). In our opinion, a good deal of this criticism stems from a failure to distinguish structural and processual concerns from content and motivational questions. When judiciously interpreted, the structural characteristics of Kohlberg's position, and the understanding of *processes* in cognitive development it includes (but perhaps not of content), and the results based upon these can be quite useful.

The aspects of cognitive development addressed by Piaget are necessary, but not sufficient for development in the domain Kohlberg considers. Thus, moral development appears to be contingent upon prior development in conceptual functioning (Sullivan et al. 1970). Kohlberg identifies three major developmental periods—preconventional, conventional, and postconventional—each having two substages. The supposed universality of this developmental sequence has been subjected to much cross-cultural research, and its major outlines at least have received empirical support (Langer

1969). As was the case for operational thought, development within a given population tends to be uneven, not all individuals in all societies complete the entire developmental sequence (Kohlberg and Gilligan 1972).

The three approaches outlined here constitute a class of theories of intellectual development distinguished by an emphasis on the formal, structural aspects of thought. Each sees development as passing through a fixed sequence of stages. All assume the innate activity of living organisms, and all are interactionist theories, rather than theories that favor the influence of either the environment or biological maturation. These theories are concerned with describing how thought is organized and with the progressive development in individuals of that organization. Finally, all three see cognition as continuous with other biological functions.

These approaches have important implications for the conduct of social science research. Here we outline some of these, focusing mainly on anthropological research. Most of the points raised in the following sections of this chapter can be easily applied to sociology, psychology, and other social and behavioral sciences.

Implications for Social Science Research

Problems of anthropological field research involve the relatively limited time period of the research, and reliance on various forms of observation and communication. Failure to understand the structural differences in individuals and institutions, and the differential functioning of structures under various environmental conditions, can lead to the erroneous representation and interpretation of ethnographic information. A sample of these issues is discussed here.

Communication

When anthropologists enter the field, they should be aware of a number of sources of bias which can affect the nature of the information eventually recorded and reported. Consider first the use of informants in the field in relation to research done in the structural-developmental tradition discussed above. Eliot Turiel (1966) assessed the complexity of children's moral-social reasoning using the dimensions described by Kohlberg. Subsequently, he presented them with moral arguments that were either one stage below, one stage above, or two stages above their level of moral-social reasoning. Children presented arguments below their own level understood these arguments but dismissed them as silly and childish. Arguments one level above their own were seen as superior to those they currently used. Arguments two levels above the child's were generally not understood. What happened in this case was that the children would reinterpret the more complex arguments in terms of their own level of organization, often drastically altering the meaning of the argument. This we call the "Turiel effect."

In anthropological fieldwork, informants serve as a conduit of information to the researcher. The possibilities for systematic bias and selection in this information are important to consider. Turiel's findings are suggestive. Individuals interpret the operational environment in terms that are constrained by their own level of intellectual-affective functioning. When questioned concerning the customs of a society, which, for example, may be postconventionally structured, a conventional informant might well transmit information to the researcher at the "lower" level. For example, when asked to interpret the United States Constitution—a postconventional institution that considers human behavior in terms of a social contract arrived at by mutual examination and agreement—a preconventional individual would describe a document containing a prescriptive, fixed, and authoritarian set of rules. Dozens of psychological studies have presented the Constitution to United States citizens as an attitude questionnaire. Invariably, only a small number of those sampled interpret the Constitution in a postconventional manner.

The informants' cognitive organization, on the one hand, introduces systematic bias into field data. The consequence can be an ethnological description that presents conceptual material and institutions in terms far more simple than is the case. On the other hand, this problem extends to the interpretations and observations of the field researcher as well. An anthropologist functioning, for example, at the concrete operational level because of stress in the field, collecting data on folk classifications and formal semantics among a group in which there are individuals who operate upon their environment at a formal operational level, might well provide a vastly oversimplified description of that society. Such a description might be a list of inclusive/exclusive categories related to each other through only one dimension. In fact, a number of the group will relate these categories to each other in a variety of ways, depending upon the perspective they adopt. Analogously, a conventional anthropologist interpreting a society with many postconventional members might well report that the society contains only two types of members: rule followers and deviants. A more accurate description would note the range of behavioral variety, ranked according to the complexity with which representative individuals operate on their social and moral environments.

The potential for the introduction of systematic biases into the results of fieldwork because of differing levels of structural-developmental functioning by informants and researchers describes the general case wherein information given and received and interpreted is biased at two points in the exchange. In a slightly different context, Pinxten (1981:59) called this a "double-bias situation." His claim is that such double-bias situations are characteristic of all anthropological fieldwork. He argues:

> An ethnographer can never be an objective outsider nor can he be a subjective insider in a particular culture, since (to different degrees) he will always be in a double-bias situation: he is biased by his own

cultural outlook and he is accepted in a certain role through the bias of the cultural group he is visiting. (Pinxten 1981:59)

The general problem could not be overcome by assigning only the most complex anthropologists to the task of fieldwork. This is because the Turiel effect not only relates to the researcher's dispositional level of complexity, which may be sufficiently high, but also to his or her functioning at the time data are gathered. Cognitive organizational complexity is a capacity that may or may not be realized in performance. The realization of this dispositional capacity will depend upon the nature of the environmental context in which the ethnographer works.

Situational variables, such as stimulus amount, complexity, or novelty, have been demonstrated to cause a decrease in cognitive functioning (recall the discussion earlier in this chapter, and see Schroder et al. 1967; Tart 1975; Rubinstein 1979a). The emotional state or degree of involvement in a situation has a similar effect (Coffman 1971). An anthropologist entering a new field setting who is anxious and uncertain is likely to systematically simplify his or her observations to note only more salient or novel aspects (e.g., see Hill 1974). The trauma often accompanying the early periods of fieldwork, commonly called "culture shock," and that upon returning to one's own society can be understood as the effects of stress upon the researcher's conceptual systems. It is important to be aware of the bias that may thus be introduced into data collection (especially by participant observation) during such a period.

During such periods of stress, behaviors and institutions will be seen as simpler, more authoritarian or hierarchical, and less sophisticated than is actually the case. There would be also a tendency to "see" phenomena in terms of previously developed expectations and to assimilate information to existing models of the environment. This helps account for the common reports of ethnographers that as they stay longer in the field the amount and quality of data collected increases. As the stress associated with the novelty of the fieldwork setting subsides, cognitive functioning probably matches more closely the ethnographer's dispositional level of cognitive-developmental organization. The improvement in data acquisition reflects this. This highlights the importance of lengthy field stays and suggests that it is useful to enter the field with a formal methodology for collecting data which may be carried out whether or not the fieldworker experiences high levels of stress.

Synchronic Study

For many practical reasons anthropology has emphasized a synchronic approach to its subject. To a considerable degree the validity of this strategy rests upon assumptions that various characteristics of a society and of social action are relatively constant over time and environmental variation. This is particularly true of research conducted within the structuralist vein (e.g., Lévi-Strauss 1969; Rossi 1974). Research on psychological structures, including the adaptive behavior of small groups, and on societies undergoing environmental change suggests that such assumptions are erroneous (Plog 1974; Count 1973).

Societies, like cognitive structures, are active systems. The formal aspects of social institutions are influenced over time by the environment in which they function. When environmental changes lead to changes in the cognitive functioning of individuals in the society, its social institutions are often modified as a result. Environmental stress, such as resource deprivation, can overload a social system and its institutions just as it does their cognitive counterparts. Internal dysfunction within a social structure can stress the system in a similar fashion. In both cases, a concretization of the structure marked by greater rigidity, traditionalism, hierarchicalization, and authoritarianism is expected (Laughlin and Brady 1978).

The synchronic study of a society or institution is open to errors in measurement due to environmental stress. A society observed in times of plenty may show a structural organization and patterning of behavior considerably different from what might be observed in times of scarcity. Similarly, observations taken during a time of stress that concretizes a society's performance will give a skewed picture of that group and of its capabilities. Societies like the Ik (Turnbull 1972, 1978) and the So (Laughlin 1974, 1978), which appear very atomized or hierarchical when observed under stress situations, may appear very different under different conditions. What is important for the social scientist is the fact that these differences may be predictable in their course and structure (Laughlin and Brady 1978; Rubinstein 1975; Laughlin 1974). Work with other primates confirms that structures of social action may vary from place to place and from time to time, depending upon demographic and ecological variables (Sade 1972; Denham 1971; Teleki 1973; Count 1973).

Effect of Cognition on Social Action

Another concern for anthropological research is the possible constraints placed upon social action by differential cognitive functioning among members of a society. Not everyone in every society reaches the most highly structured levels of cognitive functioning identified in humans. Some societies may have few members who operate at such higher levels. Research suggests that the conceptual levels of individuals in a group limit its adaptive flexibility and influence group behavior (Haan et al. 1968).

The rate and ultimate operational level of individual development within a society can constrain social development in areas of technology, social institution, and ethical sophistication. In that case, extreme cultural relativism may well be ill-founded. Any society that lacks a sufficient number of formal operational individuals may not be able to develop beyond a fairly simple technology.

It is entirely possible that the entrepreneurship requisite to economic modernization and innovation requires differentially high conceptual development in a number of group members (Belshaw 1965; Barnett 1953). Societies where optimal development does not occur may produce few entrepreneurs.

Social Intervention

How best to assist a society or group in trouble is a question often asked by concerned individuals and governments. Education or training that presupposes formal operational ability is sure to fail if that ability is not present. Social systems under stress may suffer structural damage if stress is too severe or too long in duration. Interventions designed to alleviate material stress alone may create only dependence unless provision is also made for assisting people in reconstructing the damaged structure. Cognitive deficits may derive from serious neurophysiological damage that may, in turn, be caused by chronic prenatal malnutrition:

> Thus the evidence indicates the negative effect of prenatal malnutrition on brain development and on cognitive faculties. It suggests that part of the damage is long lasting and may even be permanent. But whether or not it is permanent is not as important as the fact that it persists at least long enough to interfere with learning during the critical, early years and thus interferes with the adaptation of the child to society. Not only does he bring a deficient mind to the task of growing up, but he is also condemned to do poorly in school, with all the consequences that entails for him, his family, and the society at large. (Schneour 1974:52)

There exists tentative evidence that the effects of prenatal malnutrition on one generation may be transmitted to offspring of that generation, even though the second generation has not experienced prenatal malnutrition (Zamenhoff et al. 1971). Thus a society confronted with chronic malnutrition may not develop sufficiently complex conceptual systems among its members to comprehend and profit by interventions structured at too high a level (see d'Aquili and Mihalik 1977).

Variables influencing the functioning of human cognitive systems have important but often subtle implications for social science research. Stage development as well as environmental complexity may not only constrain the nature of information elicited from informants, but may also constrain the perception, transcription, and interpretation of data by researchers. These issues reappear in the discussion of paradigmatic science in the next chapter.

Cognitive Development, Logic and Theory

A multivariable view of cognitive development has implications for the study of logic and of the process of theory construction. The distinction between operational and cognized logics helps clarify the developmental nature of various aspects of logico-mathematical thought. Piaget and his associates examined this development in a series of experiments. Among the attributes they studied are the elementary structures of mathematics, perception, causality, and chance.

We do not here describe the results of these studies—excellent summaries have been presented by Langer (1969) and Flavell (1963)—but present a brief discussion of logic and theory in relation to this work.

In the preoperational stage, an individual distinguishes between events that are more-or-less expected. He or she is not yet capable of efficient prediction. Hence the developing system does not distinguish between those events which have been anticipated and those which were not. With the development of more highly structured operational thought, the child becomes capable of ordering the world into those events that are predictable and those that are not. Events, however, remain concrete in relation to the operations that anticipate them. It is only with the development of formal operational thought that the notion of "chance" as probability estimates develops. And, it is only at this level of abstract thought that stochastic models of the world can be created, thus allowing an individual to change her or his view of the world from that of an unpredictable environment to that of a more-or-less predictable one.

An adequate cognized logic is more than a model of human cognition. It is also a model of the development of cognition in the individual. The fact that not all people, nor all practicing scientists, function at the most abstract levels of thought at all times has important ramifications for the study of the history and practice of science.

We emphasize that the complexity of cognition underlying behavior may be difficult to infer from that behavior (see Chapter 1). The "same behavior" displayed by two different people may have different underlying cognitive structures. For this reason the social and behavioral sciences need to pay close attention to the relationship between cognition and behavior. This is particularly so since it may well be impossible, as it was for Feldman and her associates (1974), to find people capable of high levels of performance in nonverbal contexts who are able to reconstruct the processes by which they carried out that performance.

This is important in attempts to reconstruct the thought behind cultural phenomena without painstaking methodologies designed to factor out the probable structural facilities involved. Attempts to reconstruct the nature of cognition involved in myth, for example, have systematically failed to recognize this important point. Inevitably, such reconstruction results in a model of "the" cognitive structure generating the myth. Yet from a close examination of the "principles of reason" adduced by Lévi-Strauss, for example, two things are underscored by the view presented here. Virtually all of the principles (and their products in myth) may be the outcome of what Piaget termed "transductive logic," which is the hallmark of the logic performed by preoperational thought. These principles can of course also be generated by cognition characteristic of higher levels of functioning since higher level structures incorporate the lower level structures. Thus, carrying out a Lévi-Straussian myth analysis may only tell the lowest level of cognitive structural development required for the behavior or expressive form examined.

Also, most discussions about the possible differences between "primitive" and "modern" modes of thought do not control for either the distinction between structure and content, or for the multivariable nature of cognitive development. Cross-cultural studies of Piagetian cognitive development indicate that there may be cultural variation in the extent and complexity of cognitive development (Dasen 1977). However, this variation cannot be explained by reference to notions of primitiveness or modernity. Rather, it appears to be due to environmental variations which are either conducive to or inhibitory of optimal development. Further, in cross-cultural research it is necessary to distinguish clearly between structure and content. Apparent cultural differences in thought revealed by particular studies may well be restricted to the level of content and thus not exist at the structural level.

Four

Cognition and Paradigmatic Science

In 1962 Thomas Kuhn proposed that a science progresses through periodic revolutions, rather than through the gradual building-up of knowledge. He argued that a science is organized around a "paradigm" and that development in a science occurs when one paradigm is relinquished and an alternative is accepted. This view has become popular among behavioral scientists. Many philosophers of science, however, have objected to it in general and have been critical of its central use of the notion of paradigm in particular. Criticisms focus on the apparently untestable nature of the description of paradigms (Shapere 1964) and on the overly vague formulation of the construct (Masterman 1970). In response to this criticism, Kuhn has in several places elaborated his notion of paradigmatic science, and he has developed more fully his view that a science passes through periods of "normal" activity punctuated by revolution.

Despite the philosophical criticisms that have been raised against Kuhn's account of development in science, it is our view that it is substantially correct. The source of the apparent difficulties in Kuhn's original presentation results from its collapsing together many levels of analysis. In his later work, Kuhn addresses this difficulty, and he begins to specify more fully the different levels at which paradigms work. Still lacking, however, is a sense of the processes underlying the functioning of paradigms. We believe that viewing them from the perspective of science as cognitive process helps both to explicate the notion and to demonstrate its robustness.

This chapter, then, sets out a view of paradigmatic science as cognitive process.

The Nature of Paradigms

Kuhn noted that science rarely functions in a purely inductive mode. Rather, scientists begin with an accepted model or pattern of nature which acts like accepted judicial decisions. These are the "common law" of the scientific community. This model limits how a community of scientists defines the domain in which it is interested, identifies the phenomena appropriate for its observation, and legitimates a set of methods for investigating those phenomena.

Typically, a paradigm is accepted by a scientific community because it both accounts for phenomena of interest (some of which have until then resisted explanation) and because of its promise for solving other outstanding problems. A paradigm leaves many questions unanswered and therefore frames a science's activities for some time after its acceptance. The kinds of problems that remain, however, are of a fairly pedestrian nature, and the research activities that are needed to resolve them Kuhn likens to "puzzle-solving." It is this kind of work that takes up the time of most scientists and of most sciences. These periods of relative calm Kuhn calls "normal science," and he argues that they are characterized by the more-or-less straightforward application to small problems of the methods of investigation legitimized by the paradigm.

In the periods of normal science the fundamental assumptions of a discipline are accepted without question. The application of fairly standard methods of investigation to small problems yields many solutions but also some failures. Problems for which solutions are not found are either subjected to continued scrutiny or put aside as out of the range of solution by available methods. The core of the paradigm is not as a result itself subject to examination.

In fact Kuhn argues that the basics of the paradigm are so fully taken for granted that they influence not only the activities of members of a scientific community but also how they view the world. Scientists working within traditions defined by different paradigms actually "see" different things when looking at the same aspects of the world. The result of this is that even if they use the same technical terminology to describe what they see, observers committed to different paradigms will mean different things by those words. Thus, a shift in the paradigm to which one is committed is much like a shift in one's world view.

Recently Kuhn (1970) identified two levels upon which paradigms function. On the social, or sociological, level, paradigms may be seen as the world view of a given scientific community. "A paradigm governs not a subject matter but rather a group of practitioners" (1970:180). Such a community may be defined in various ways: by discipline, by membership in a specialized area of work, by explicit commitment to an approach to investigating a significant problem, and so on. At this level, paradigms refer to group commitments. These mutual commitments combine to form a "disciplinary matrix." This matrix includes (1) symbolic generalizations such as formulae applied to the study of a class of phenomena, (2) shared beliefs in particular models which legitimate the use of certain analogies and metaphors, (3) common values about what is important, and (4) concrete problem solutions (called exemplars) so firmly accepted that they constitute a critical aspect of the training of new scientists.

The second level at which a paradigm exists is in the conceptual systems of individual scientists. To one degree or another, each component of the disciplinary matrix is part of the cognized environment of each practicing scientist, just as it is a group characteristic. At first this claim appears contrary to the necessary existence of intracultural variation we described earlier. It is not surprising, therefore, that it received much sharply negative attention. Understanding the general relationship of cognitive systems to behavior helps resolve this apparent contradiction.

The way in which paradigms exert control over a scientific community is in part determined by general processes (outlined in Chapters 2 and 3). It is further restricted by the special characteristics of the cognitive processes specialized for use in coordinating group behavior for corporate activity. It is this nonordinary set of processes which characterizes religious ritual, for example, and which we believe accounts in large part for the elimination of significant intracultural variation within a scientific discipline. This discussion of paradigmatic science highlights the similarities between ritual and science, and at the chapter's end explicit parallels are drawn.

The relative lack of intracultural variation in scientific communities is revealed by the actions and communications of scientists. Nowhere is this clearer than in the training of young scientists. During this socialization process the paradigm is defined for the initiate through the exemplars of the disciplinary matrix which he or she seeks to enter. It is in the context of scientific training that aspects of the paradigm become part of individual experience. During this period the exemplars are constituted as actions, and they are encountered in published models. The practice problems, exercises, and laboratory experiments with which each initiate must successfully deal constitute his or her experience of the paradigm. It is out of these activities, this set of operations, that the paradigm-defined view of the world is internalized and comes to constrain what each scientist "sees."

A Multilevel View of Paradigmatic Science

A science may be viewed as passing through three phases: (1) a period during which the paradigm ultimately accepted by the scientific community is constructed—

the pre-paradigmatic phase; (2) a phase of relative calm during which the puzzle-solving activities described by Kuhn are carried out—normal science; and (3) a period in which an accepted paradigm becomes unstable and alternative views gain adherents. It is only in the second phase, where views different from those endorsed by the paradigm are discounted and in which the paradigm is assumed to be "true," that the productive work of a mature science is carried out.

It is generally assumed that paradigms are correct. They are seen as the proper and the only productive way to view the scientific domain with which they deal. The scientist's activities are expressions of a paradigm; these are defined by it and constrained by it. Implicit in these actions is the expectation that the paradigm will be confirmed. The paradigm itself admits no disconfirmation—only errors in meaning, data collection, or analysis. As such, disconfirmations are generally not fed back into the core of the model but lead rather to slight adjustment of the paradigm, thus bringing about only minor modifications in patterns of scientific behavior. Normal science is something of a closed loop, minimally, if ever, open to serious modification at the core. Scientific behavior serves to maintain and solidify a paradigm's dominance in addition to defining how problems are solved.

The acceptance of a paradigm by a scientific community stabilizes the field, and it directs communal energy toward the solution of common problems. The problem-solving efficiency of the science is then at its peak. Foundations and first principles are taken for granted, and their essentially hypothetical nature is forgotten. The professional activities of scientists become constrained to a form and range prescribed by the paradigm. Thus, at any particular time a paradigm characterizes a stage in the development of a science. The view of a paradigm as a phase in the discontinuous development of a science can only be gotten by taking a perspective other than that of the working scientist. Because paradigm shifts involve the reinterpretation of earlier paradigms, the working scientist sees a history of continuous development toward even more accurate models of the world.

When paradigm-constrained behavior fails to produce solutions for research problems, anomalies result. Contrary to what might be supposed, these do not generally threaten the integrity of the paradigm unless (1) the anomaly is deemed critical or (2) these anomalies become too numerous. In both cases the paradigm is fiercely defended (and not always by methods that fall within the bounds of scientific decorum; see, e.g., Koestler's [1971] discussion in *The Case of the Midwife Toad*) until a viable contender is presented. Even then it is generally younger members of a field who adopt the alternative view that eventually replaces the old paradigm. Proponents of the old paradigm are often unable to reconcile their past experience with the new paradigm and are generally unable to "see" in the new way. On both levels, paradigms can be separated into two poles of a cybernetic circuit. One pole is the model, the paradigm itself. In this sense the paradigm is part of a belief system. It is the expectations and modes of conceptualizing the world of each scientist and those common to the scientific community. At the other pole is the individual and corporate behavior through which the scientific community seeks to describe and explain phenomena in ways consistent with the paradigm. Behavior serves to construct and to maintain a paradigm.

This behavior plays a critical role in the recreation of the paradigm in each new member of the field. The individual and group experiences of each new generation of scientists, with the behavioral working out of the paradigm—bit by bit, problem by problem—recreates the paradigm. The paradigm is transmitted in each new generation by the perception of analogies in problem-solving, by the creation and habituation of patterns of actions coordinated with perception. The coordination of perceived elements of a (paradigm-defined) problem and a repeated range of paradigm-approved operations and actions produces a habituated perceptual-motor pattern that is experienced as that science itself. This training sequence of prescribed activity together with textbook-presented fundamental concepts and rules of action produces an adaptive mode with cognitive, perceptual, and behavioral components that becomes relatively taken for granted. The reasons for believing, seeing, or doing this science are not questioned (Lomnitz and Fortes 1981; Lomnitz and Lomnitz 1977).

Paradigms and the Issue of Anomalies

Perhaps the most critical aspect of Kuhn's view of scientific progress is the notion that paradigms have a serious effect upon what scientists actually see. This is an easy issue to overstate. Do paradigms determine what scientists actually see? Is "constrain" a better way to characterize the process? Perhaps even "predispose" is more correct? Settling upon a way of characterizing the effect of paradigms is important. The issues involved must be examined, and the processes involved clarified. (In the following discussion we use "seeing" metaphorically to mean "understanding" or "recognizing the implications of." This use follows that of Hanson [1958], Kuhn [1962] and others and is important for connecting our discussion to work in that philosophical tradition. However, our aim in this chapter is to move away from such telegraphic discussion and to specify more fully the processes implied by this way of using "seeing." Thus, although we begin by using this terminology, our goal is to move beyond it.)

What exactly does it mean "to see" when we refer to a scientist (or even a science) seeing a phenomenon? Kuhn says:

> Surveying the rich experimental literature from which these examples are drawn makes one suspect that something like a paradigm is prerequisite to perception itself. What a man sees depends upon what he

looks at and also upon what his previous visual-conceptual experience has taught him to see. (1970:113)

Thus Kuhn attempts to ground his exposition on certain experimental findings of cognitive psychology. For instance, he presents experimental data to demonstrate how perceptions are determined by prior experience. Drawing on the early work of Bruner and Postman, he mentions that subjects felt distinctly uncomfortable when presented with playing cards on which the colors of some of the suits had been reversed (i.e., spades and clubs were red, and hearts and diamonds were black). Subjects would make identifications as though no anomalous stimuli were presented. Even more important for Kuhn's position, the subjects were unable to identify the source of their discomfort until exposure to the anomalous cards was increased well beyond the time necessary for ordinary pattern recognition. Kuhn concludes from this evidence that under normal conditions subjects are unable to detect anomalies, that their past experience with normal decks of playing cards determines how they perceive the experimental deck.

Is it true that because subjects were unable to verbally identify the source of their discomfort that they were unable "to see" the anomalies? If they cannot "see," why the discomfort? In the routine execution of the sorting task, early models appear to determine seeing; the sorting process goes on automatically (like normal science), and the sorting is dependent upon one or the other characteristic dominant for the sorter. With closer inspection of the data an alternative model is settled upon. This is not so much a dichotomy as a range of functioning in seeing. Kuhn rightly points out the relative incomparability of such experiments with highly precise scientific observation, yet it is also true that the psychological investment of scientists in the validity of the paradigm in which they work may contribute a still more powerful bias *in favor* of misperception. Something very intriguing happened in the playing-card experiment. Something still more intriguing occurs when scientists display similar behavior.

Psychology contains an interesting example of this. Though pre-paradigmatic in Kuhn's sense, psychology has tended to behave as if consensus has reigned these last fifty years. The dominant approach has been one or another form of behaviorism. Seligman discusses the fate of one anomaly in this field, a study by Garcia and Koelling. This research produced startling results incompatible with strict behaviorist premises. It demonstrated that, when a couple of generally accepted parameters for research were tested, results of conditioning experiments were contrary to what would be expected on behaviorist premises. Important here is neither the particulars of the research nor its results, but that the study was unable to enter the major professional literature for many years. This study was not published because "it couldn't be accurate." Once recognized as a valuable study, however, the door was open to a large number of similarly anomalous studies, all of which were "not possible" under the assumptions of the earlier viewpoint of the majority of academic psychologists. They were unable to see anomaly *qua* anomaly; unable, that is, to see the anomalous results *in the same terms* as expected results. Garcia and Koelling had looked at parameters foreign to the accepted experimental situation and saw results simply not permissible within the views of the community of psychologists. The results, however, were there, were real, and were experimentally sound.

Of course, historical instances of this kind are available in other sciences as well. For example, it is clear from Arthur Koestler's (1971) account of the controversy in the early 1900s between neo-Darwinians and neo-Lamarckians, centering on Paul Kammerer's work on the nature of inheritance, that the defenders of the dominant view (neo-Darwinism) of inheritance ignored at times evidence supporting the opposition's position and at other times "saw" the evidence but did not act on it. The controversy surrounding Kammerer became particularly bitter as it developed, and when it was discovered that one of his important specimens had been doctored, Kammerer's work fell rapidly into complete disrepute. Yet, Koestler argues, the available evidence suggests that most of Kammerer's empirical work remains unchallenged, though it too was taken to be discredited.

These examples are not too far from the playing-card problem. What did the journal editors "see" when they read the Garcia and Koelling study and looked at its statistics and graphs? What were they aware of and how did they perceive this and similar anomalies? A study was conducted, analyzed, and written up within the general canons accepted by psychological researchers. This information was submitted and read, seen by the most competent scientists. We do not really know what effect this stimulus had upon each individual cognitive system exposed to it. We do know that it did not enter the critical social system of psychological knowledge, and we know that it was anomalous to the position supported by studies accepted into that system. Why was it not assimilated into the general fund of knowledge? What were its individual cognitive and affective consequences?

Interesting experimental data suggest that this sort of effect is not an odd occurrence among journal referees. Using the pretense that he was assembling an anthology, Mahoney (1976; Mahoney et al. 1978) circulated to seventy-five reviewers for a "well respected periodical" a number of manuscripts dealing with controversial subjects in behavior modification. The reviewers were instructed to use the same editorial criteria they did when reviewing for the journal. He summarizes the findings as follows:

1. When reviewers read manuscripts in which the data supported their presumed perspective, they rated its methodology as "adequate" or "excellent," and recommended that it be published.

2. When the same procedures yielded negative results, they were rated as "inadequate," and the reviewers recommended rejection of the manuscript.
3. Mixed results manuscripts were not well received, regardless of how they were interpreted.
4. Manuscripts from which the results had been deleted were favorably reviewed.
5. Reviewers showed very poor reliability; their average agreement with one another ranged from 0.30–0.07 across various dimensions of the manuscript. A subsequent study used another sample of reviewers from two other technical journals. The manipulated factors were the author's institutional affiliation and the presence of self-citation; all manuscripts were otherwise identical. In half the manuscripts, three of the references were cited as "in press" by the hypothetical author; in the others, these same "in press" references were credited to another person. Their recommendations suggested that referees were not influenced by affiliation, but were significantly kinder to an author who could cite his own "in press" publications. (Mahoney 1979:358–59)

With the advantage of hindsight we now know that this pattern is common. What accounts for this pattern of behavior?

We presume that these people "saw" such research as unacceptable to their science and its progress. But, what did they see and what did they not see? Uric Neisser (1976) points out that perception is an active, adaptive process. He notes, "When perception is treated as something we do rather than something thrust upon us, no internal mechanisms of selection are required at all" (1976:84). The tendency is to account for not seeing by positing some sort of filter on perception. In the current context, for example, paradigms would be thought to serve as an inhibitory mechanism keeping the natural flow of information out of the cognitive-social systems, rather than as a critical adaptive-system that seeks out information for inclusion into itself. Neisser's thesis is supported by studies using visually overlapped, televised games where subjects are told to attend to one and ignore the other image. Results indicate that this is extraordinarily easy to do. One can follow and use a given primary visual event and ignore another equally present to the eyes. "One does not see the irrelevant game" (Neisser 1976:86). Neisser continues: "Only the attended episode is involved in the cycle of anticipations, explorations, and information pickup; therefore only it is seen. Attention is nothing but perception: we choose what we will see by anticipating the structured information it will provide" (1976:87).

Neisser used a model of the perceptual cycle consisting of processes extended in time and location within a neural system. That part of the system which he calls a schema is that part of the perceptual cycle internal to the perceiver and modifiable by experience. Perception is a form of behavior predicated on an internal model. Such a model has system specific sensitivities or expectations of what is "out there." From this point of view, perception is an information-seeking process that "expects" to find a certain range of information and selectively disregards information not in that range. *The general point is that rigid paradigms make the sciencing enterprise overly assimilative at the expense of accommodation, and this results in a loss of adaptive balance.*

Organic perceptual systems may actively seek out information within the definable limits of their cognitively and neurologically constructed models. Such information is perceived and assimilated into their functional cycle. This does not mean they react only to information that confirms what is expected, only that they *attend to* and use such data. Such systems treat unexpected input as the null case.

This is very different from seeing the evidence for, or as it would be defined by, a different point of view.

Three classes of information are potentially perceivable by an active perceptual system: (1) information as internally defined by the system; (2) information different in kind from that anticipated by the system; and (3) information contrary to expectations though not necessarily different in kind. We are faced with a system that is binary in nature, a system that picks up confirming signals and disconfirming signals. Such a perceptual system functionally collapses possibilities 2 and 3 into a class nearly identical to 3. Disconfirming experiences involving the speed of light were perceived by physicists as problems, as anomalies, as not confirming the expectations of classical mechanics. It took Einstein to see the constancy of the speed of light as a principle. Everybody was looking at the same phenomena, but where the others saw disconcerting anomalies within an expected reality that was correct, Einstein saw the phenomena *qua* phenomena, and with it an entirely different reality. Seeing an error message in a stable, predictable world is quite different from seeing the same message as indicative of an entirely different world.

Paradigms and Cognition

The simple model just sketched describes an adaptive, information-seeking system. This system has two poles: (1) a central representation, or functional organization; and (2) a system of behavior capable of adaptive interaction with the world. This second pole includes behavior in the coordinated, motoric sense and in the active, perceptual sense. It is biased toward conserving that model or organization by limiting its operations with the environment to those permissible within the model and by perceiving aspects of the environment selectively. The capacity to notice anomalies derives from operations in the model-environment interaction, and from the perceptual counterpart's ability to find the sensory correlate. Note that the perceptual system can be seen as an integral aspect of a perceptual-motor system, constituting the interface between the paradigm and the world it seeks to model. The principal work of this behavior system is to bring the world into concert with the model or paradigm.

Since paradigms are the main mechanism by which we bring our understanding of the world into concert with the world itself, this may seem a paradox. Yet paradigms are assumed to be, in some sense, a true representation of the world, needing only local modification. If perception is really the action based on anticipations defined by the paradigm, we seem to have a *cul de sac*. This problem arises only when science in general and paradigms in particular are viewed ahistorically. It is Kuhn's thesis that revolutions change not only paradigms but also how events are perceived.

Parallel phenomena to the relationship between paradigms and perception, and to the issue of anomalies, occur when we observe the development of cognition in children. Piaget (1970) sees the child in the role of a naive scientist, developing his or her view of reality over three major periods or states, each subject to revolution, displacement, and incorporation into a new view of reality. How the physical world is perceived and what is perceived changes as a function of these revolutions in ontogenetic development. The characteristics of perception and its relation to the cognitive system are roughly the same as described for paradigms. As with paradigms, each period can be viewed synchronically or diachronically, and in the latter case seem to pass through similar stages of preparation, consolidation, and disruption.

In the case of cognition this progression forms an invariant sequence, the succeeding stage displacing the former over time through repeated encounters with anomalies or negative feedback until a critical level either in cumulative quantity or in critical areas is reached. Changes in perception develop in parallel yet appear contingent upon development of cognitive structures. Roughly at the ages of five through eight, children pass through the preparatory phase and into the stable phase of the second of the three stages of development (see Chapter 3). A remarkable shift in their perceptions of the world takes place during this period. The progression is from a mode of acting on the world in ways limited to perceptual-motor coordinations to a mode where these coordinations are internalized mental operations.

Inhelder, Sinclair, and Bovet (1974) examined this development longitudinally with preoperational and concrete operational children as well as with some children actively in transition. Some of their experiments had to do with conservation of physical properties of matter: rolling the same amount of clay from a ball into a sausage, pouring water from a tall, narrow container into a short, wide one, and similar exercises. Concrete operational children easily note that mass and volume remain the same through the transformation—the same amount of clay is in the ball as in the sausage. The preoperational children do not; they see only one dimension at a time—height or width in the pouring problem, for instance.

When asked, the younger children say that quantity or volume changes with the pouring, rolling, etc.: "There's more, it's tall" or "There's less, it's thin." They appear to attend to one or the other dimension in their judgments, focusing on the static properties of the beginning and ending states, missing the transformation. If thoroughly questioned, these children often reveal that they saw, or noticed, that the taller jar of liquid is thin. Yet it does not enter into their judgments. Even when, in some sense, perceived, they are not "seen" in an effective way. Things are either tall *or* thin, short *or* fat. When both are noticed it is in sequence. The perceptions are not linked or coordinated. Preoperational thought does not permit such coordinated perceptions. The information is not picked up by the perceptual system. Perception is anticipation, and a preoperational system anticipates static properties rather than transformation, single attributes rather than coordinated ones. The child's visual system registers other dimensions, but he or she "sees" only one. The anomaly is noticed but not "perceived."

More information is gained by observing children in transition between stages. Their judgments often fluctuate wildly. Sometimes they will see the transformation. Change in seeing is a function of an altered internal model. This is closer to the scientist than Bruner and Postman's card experiment—the child as natural scientist who sees but does not "see" anomalies. The child's perception changes as a function of the properties of internal structure, as a function of the child's natural paradigm. We specifically note the oscillation in judgment of stage-transitional youngsters, who already have an alternative model at hand.

Once fully achieved, the new mode of seeing concrete operationally is firmly set. The new structure supersedes the old, including the old elements in the new, more comprehensive system. It becomes automatic and guides perception until the next developmental shift around eleven or twelve. The child then sees transformations and is no longer limited to seeing successive states. The world changes drastically. For years the child repeatedly applies the new view to problem after problem, doing the normal science of a concrete operational child. This metaphor is not a flippant one. The child is a scientist, discovering laws of physical reality as did earlier scientists, and acts on these, even if he or she can not describe the principle.

Does this analogy go too far, or not far enough? Here is just an overview of points more fully developed elsewhere and in different contexts. How closely does Kuhn's notion of paradigm resemble our model of cognition, and what light can be thrown from one onto the other?

Paradigms, like cognitive structures, are characterized by a relative suddenness of emergence and then are subsequently applied to classes of phenomena over time. For example, concrete operations are progressively applied to notions of conservation of quantity, mass and volume, causality, and other areas of experience. When looked at diachronically, cognitive structures display the same three general phases—construction, stable equilibrium, and disintegration—prior to the emergence of more complex structures. In the stable state, they relate to behavior as paradigms relate to "normal science."

Behavior stands in the same relation to cognitive structures as it does to paradigms. That is, cognitive structures are built up from perceptual-behavioral interaction with phenomena and the abstraction and internalization of common patterns of interaction or schema, rather than aspects of the phenomena per se. When the structure is completed, behavior functions to confirm the model and apply itself to an increasing range of experience. A change in perception accompanies this extension. The world is "seen" in different terms; the same sensory reality is perceived differently, the same physical properties are restructured into new perceptions.

Anomalies also are generated and treated the same way for both cognitive structures and scientific paradigms. Individual incidents of nonconfirmation have little effect, until *patterns* of such disconfirmation become evident. In critical cases, discomfort is evident in both instances, and at critical times of transition, uncertainty increases leading to a questioning of basic premises. In both the developing child and the paradigm at the end of its usefulness, the question becomes one of a new way of seeing. In both cases, the old is succeeded only by a way perceived as "better." In each case, disconfirmation of the accepted model is seen not as an incidence of another model but as a disappointment of the existing one, a technical error with slight difference.

Paradigms can be construed as existing at both the level of discipline and the level of individual scientists. Paradigm shifts generally happen at the level of the discipline. Scientists who do change accept the new paradigm; those who do not are replaced by new scientists entering the field. This perspective is strikingly close to Waddingtonian genetics. In each case the population has a reaction norm, reforming itself into new structures based upon the adaptive efficacy of individual response patterns. Individual cognitive fields may be treated as populations that restructure themselves, reorganizing their substructures into new adaptive fields, or "populations." In each realm, paradigms, genomes, or cognitive structures, we are dealing with overall reaction norms, total systems in the act of adapting and thus altering their adaptive range over time.

The present discussion sets out a research problem for philosophers of science: Examine the data from the history of science for evidence of an organized development, similar to that proposed by Piaget (1970, 1971a) for cognition, and by Waddington (1957) for the evolution of species. One question to ask, for example, is Does the progressive elaboration in sciencing of cognitive structures that seek to improve the fit between the operational and cognized environments take place in a manner analogous to that seen in the ontogenetic growth of cognition? The possibility of such a venture is suggested by the fact that succeeding paradigms generally are characterized by a key criterion for such development: stable explanation and prediction over a range of phenomena greater than that of its predecessors. The next step would be to work out internal rules of organization, transition, and progression parallel to those being developed in other fields.

Some Proposed Principles

Cognitive structures can be transformed in three ways. They can be transformed at the surface through an internal reordering of functional substructures and thus create a different pattern of similar complexity, or they can be transformed developmentally by organizing these subsystems into more complex structures. The former can be compared to cultural differences as viewed by

Lévi-Strauss, the latter to cognitive development as seen by Piaget. Each can be construed as a transformation in world view, and each suggests a different way of seeing.

These two transformations have a determining influence on perception and perceptual-behavioral systems, which was termed "sensory transformation" (see Chapter 3). A third type of transformation refers to the transforming activity of behavior (including perception) as it acts upon the operational environment. This is the transformative activity of internal structures, linked with the perceptual-motor system. We experience not the world, but the results of our interactions with the world. In the case of advanced cognitive systems, these interactions extend to internal action or operations in the Piagetian sense.

These three types of transformations form a nested, hierarchical model of cognition, the principal limits of constraint resting in the developmental range of structure. This forms the parameters of a further constrained system of surface transformations which, in turn, imposes constraints upon the functional range of the sensory transformation system. It is at this third level that interaction with the operational environment is carried out. The perceptual-motor system for any cognized environment is defined by its other two systems.

The sensory transformation system "feeds" the cognized environment; it seeks out information within the range defined by the internal organization. An optimal balance is necessary to keep the internal system active and open to development. Outside that range, two forms of imbalance can occur: (1) distortion in favor of confirming the existing model, or overassimilation, and (2) overaccommodation or distortion toward disconfirmation of the model. The former entrenches the central model; the latter prevents stability.

Earlier work applied this model to ritual behavior (d'Aquili et al. 1979; Laughlin and d'Aquili 1974). It is also useful to apply it to paradigms. In the earlier treatment of ritual it was noted that these transformations could each be viewed in a temporal sequence, similar to one observed by Piaget for development (Piaget 1971b) and by Chapple and Coon (1942) for cultural transformations such as *rites de passage*. This approach is extensively explored by d'Aquili et al. (1979). Here we note only that the principal advantage of this model is to outline a mechanism by which structures initiate departure from current patterns of functioning, "decide" on an alternative, and integrate the change into an altered structure. Two contributions are particularly important: (1) the development of a framework potentially applicable to structures at any level from the neurological to the cultural and (2) the noting of a specific step in the sequence, the alternation phase where the "decision" for stasis, surface change, or developmental change is set. In principle, the model should be applicable to the study of scientific paradigms. It addresses itself to the question of how anomalies are noticed, perceived, and taken in or rejected at any level of structure.

Such a complex, nested system allows for (1) the detecting of anomalies at levels peripheral to the level in question; (2) the admitting of data into the system which nonetheless remain out of effective range of that level as long as they do not exceed tolerance levels for disconfirmation; (3) the penetrating into adjoining levels of disconfirming signals should tolerance be exceeded; and (4) the potential restructuring of the level of focus. The key element in the overall sequence is the alternation phase of the transformational sequence and the temporal simultaneity of alternative resolutions appearing in the sequence.

The following is a sketch of this process:

A. *Genesis:* In both cognition and in scientific paradigms, the structures are constructed through behavior coordinated with perception. The general pattern of perception-action is internalized to construct an implicit perceptual system. This, in turn, effects perception of subsequent events. It creates the schema by which relevant phenomena are perceived outside the training situation. What is created is a cognitive structure in the individual scientist reflecting the viewpoint of the current culture of the scientific community. The resulting range of activities constitutes a surface structure. The evaluation of observed phenomena is constrained to the level of complexity reached by the field as a whole.

B. *Balance:* Perceptual-motor interactions within the field of study would be constrained by the limits in A. In Kuhn's words, "Operations and measurements are paradigm-determined. Science does not deal with all possible laboratory manipulations. Instead it selects those relevant to the juxtaposition of a paradigm with the immediate experience the paradigm has partially determined. As a result, scientists with different paradigms engage in different laboratory manipulations" (1970:126). Individuals with different cognitive structures engage in different operations. Both perceive the world differently. The operations and therefore the perceptions of both are constrained not only by a developmental limit but also by a "cultural" limit. Both see what they are constrained to see, and behavior is executed to confirm those expectations. Anomalies are registered in the form discussed above and enter into a hierarchical control system as suggested above.

C. *Exchange:* During "normal" functioning, information about what is seen is constrained by the organizational properties of A and especially its subset B in both cognition and scientific paradigms. During such times information based upon constrained operations is biased toward overassimilation. As anomalies accrue and critical disconfirmations are encountered, accommodations at the periphery begin to destablilize the system. Eventually these accommodations create an overall state of critical instability, or crisis, preceding revolution. Such a situation encourages regression to a pre-paradigmatic state and the reemergence of competing views.

Normal paradigmatic science and normal cognition function in a state of relative habituation analogous to the physiological use of that term. If the preceding hierarchical model is accurate, the relationship is more than analogous, it is homologous. In such a view, dishabituation could be effected either upward or downward within the system, extending from the neurological to the social. Since paradigms have both social and individual levels, dishabituation can irradiate from either level. Habituation implies lack of awareness of the assumptions that constitute those principles. By and large, Newtonian principles were unquestioned before Einstein, within the discipline of physics. The crisis state sets up the possibility of questioning, arising at this level.

While Piaget states that regressions are not seen in cognition once a stable structure is achieved, this observation is probably limited by his methods and chosen subjects. Observations of more social situations, induced environmental stress (Schroder et al. 1967; also Chapter 2), or in psychopathological states suggest otherwise (Arieti 1974). In such studies we see the following phenomena: (1) overall functioning follows an inverted U-curve function (Schroder et al. 1967; Schroder 1971) with optimal functioning matching optimal input and regression occurring when that point is surpassed; (2) observations from psychopathology suggest that this nonoptimal stress is internal as well as external to the system; (3) the overall structure may regress to a preceding level or continue to dissociate into uncoordinated substructures of differing influence on perception or behavior; (4) a new structure emerges either at the same or at a more complex level coalescing the diverse element around either a dominant substructure or one that is reformed into an entirely new mode.

The above sequence of habituation, regression, and restructuring also holds for ritual. Ritual acts to conserve the existing system by repetitive application of its principles through behavior, as a form of socialization, to protect the model against stress, and finally, following stress to reorganize the model in a new mode either intentionally, as in *rites de passage,* or spontaneously, as in conversion experiences.

Although the tendency is to see ritual in religious contexts, science too is a ritualized form of cognition. This result places the view of paradigmatic science taken here closer to Kuhn's original sociologically derived formulation of progress in science (Kuhn 1963). Further, by doing so on the basis of biopsychological considerations, the present work adds a certain robustness to the view. As Kuhn (1970: 157–58) says:

> Paradigm debates are not really about relative problem solving ability, though for good reasons they are usually couched in those terms. Instead the issue is which paradigm should, in the future, guide research on problems, many of which neither competitor can claim to solve completely. A decision between alternate ways of practicing science is called for, and, in the circumstances, that decision must be based less upon past achievement than on future promise. The man who embraces a paradigm at a new stage must often do so in defiance of the evidence provided by problem solving. He must, that is, have faith that the new paradigm will succeed with the many large problems that confront it, knowing only that the old paradigm has failed with a few. A decision of that kind can only be made on faith.

Sciencing is a class of normal cognition, evidencing the same principles in form and function as cognition in general. Further, we propose that this class of cognition is ritualized. It is more formalized in its content and expression than everyday cognition per se. The ritual nature of scientific cognition (particularly the training phase) regulates not only the interaction between the scientific community and its subject matter but also the relationship between paradigms at the discipline level with their manifestation at the individual level.

Paradigms, the cognitive structures guiding scientific behavior and perception, are ritually inculcated in the new practitioner in much the same way as religious ritual is socialized. The ritual or formalized enactment of appropriate cognitions expressed in behavior generates a limited range of perception of the environment to a fairly rigid subset of possible perceptions and provides the stability necessary for the problem-solving conduct of normal science. This allows a pragmatic mapping of the operational environment, in some sense bringing theory and reality into a degree of concert capable of supporting adaptive behavior over a wide range of phenomena.

This constraint on the individual investigator, aggregated over individuals, functions to reinforce the reigning scientific paradigm, under most conditions. One level develops and subsequently constrains perception and behavior in such a fashion as to ensure continued preservation of the paradigm itself. At both levels, awareness of the relativity of the model fades, reminiscent of neurological habituation, immersing both levels into an affirmed and unseen symbiotic bond tantamount to a *participation mystique*. Further understanding of the nature of reality is constrained and inhibited in a developmental sense, while reality is explored in a synchronic and pragmatic sense. This phase is ritually directed and controlled. This linked conceptual behavior system forces an actual constraint and standardization of "seeing" on the part of the individual in the scientific community much in the manner of religious ritual.

Five

Theory Reduction as Cognitive Process

The first two chapters of this book sketched a view of the nature of human cognitive process and suggested in a general way how this view might apply to the study of the scientific enterprise. Chapters 3 and 4 applied this view to specific aspects of sciencing at the sociological level. This and the next chapter turn to an examination of two central conceptual issues in science—theory reduction and explanation—and treat them from the perspective of science as cognitive process.

Social and behavioral scientists have recourse to a number of different "levels" of structure in their explanation of social phenomena (e.g., neurophysiological, psychological, sociological, "deep" and "surface" structure). Social theory has often utilized several of these levels, particularly those pertaining to behavior, social action, ecology, and the genetics of behavior (cf. Count 1973; Rubinstein 1979a, 1979b; Chapple 1970; Fishbein 1976; Laughlin and d'Aquili 1974; Chagnon and Irons 1979). The movement among various levels of organization raises two important questions: (1) What is the ontological-epistemological relationship between levels of systemic organization? (2) Has what is known of the nature of human cognitive process anything to offer to an understanding of the relationships between levels? Consideration of these questions necessarily involves examination of the process of theory reduction as traditionally conceived and, later, the nature of explanation in science.

Anthropology and the Philosophy of Science

One of the arguments of this book is that the most profitable approach to science, and to the analysis of the process of sciencing, is a disciplined integration of theoretical-empirical considerations and philosophical reflection. The two approaches are mutually beneficial. As Albert Einstein (1949:683–84) put it:

> The reciprocal relationship of epistemology and science is of a noteworthy kind. They are dependent upon each other. Epistemology without contact with science becomes an empty scheme. Science without epistemology is—insofar as it is thinkable at all—primitive and muddled.

Social scientists often remark that bureaucratic policy statements based upon their work reveal a lag of ten years between theory and research (Gardner 1981). Similarly, philosophers of science could well note the lag between developments in their discipline and anthropological work based upon it. For example, in their desire to create a scientific archaeology, archaeologists rushed to embrace the received view of explanation just as the philosophical community was coming to realize the inadequacy of that view (see Binford and Binford 1968; Fritz and Plog 1970; Watson et al. 1971; cf. Suppe 1977; Salmon 1971). Perhaps a more important example of discrepant developments in anthropology and the philosophy of science is offered by the common abuse by anthropological theorists of the relationship between deduction and induction. Some theorists proceed firm in the belief that the social world will order itself if we collect sufficient data. Others feel that their formulations about the world are complete if they are internally consistent and account for the data from which their formulations were drawn. The abuse of this distinction may be due, in part, to the common misunderstanding of the nature of induction and deduction. As Rudner (1966:66) put it,

> The vulgar notion that induction and deduction are "opposites" as well as its equally untutored concomitant notions that deduction is "going" from the general to the particular, and induction is "going" from the particular to the general, are not only quite mistaken, but seriously misleading. . . . There is surely no longer an excuse for repeating such bits of foolish folk-lore "logic."

Neither position is tenable in light of current philosophy of science. Moreover, from the perspective of this book, the process of scientific theory construction must be viewed as involving an alternation between deduction and induction (also Laughlin and d'Aquili 1974).

The fundamental problem in bridging from philosophy of science to anthropology is the existence of discipline boundaries. Anthropologists tend to rely upon the philosophy of science as a source of "solutions" to logical and methodological problems, rather than as a body of rational problems with which to interact. The effect has been to establish a one-way exchange between the two disciplines. Anthropologists often uncritically accept philosophical positions and introduce these into the anthropological literature via argument from authority rather than rational dialectic. One serious result is that anthropologists have held the philosophy of science at arm's length, to the detriment of both disciplines.

Over the past two and a half decades, many philosophers have realized the value of drawing empirical considerations—largely historical—into their discussions of the philosophy of science. As Suppe (1977) pointed out, the introduction of historical considerations into the philosophy of science provided the impetus for many of the recent developments within the discipline. Unfortunately, philosophers have for the most part continued to view questions pertaining to the psychological basis of sciencing as though they are irrelevant or satisfactorily answerable by means of introspection. While Reichenbach's (1938) distinction between the context of discovery and the context of justification is coming to be seen as viable only as one of analytical convenience, in practice it is rigidly maintained.

Piaget has been particularly forceful in taking philosophers of science to task for their neglect of empirical considerations. He has argued (1973:12):

> Introspection alone is not enough, because it is both incomplete (it grasps the results of mental processes and not their intimate mechanisms) and distorting (because the subject who introspects is both judge and party, which plays a considerable part in affective states, and even in the cognitive sphere where one's own philosophy is projected into the introspection).

We agree, and therefore the remainder of this discussion assumes the necessity of empirical considerations in understanding reduction and explanation in science.

The Received View of Theory Reduction

As a first approximation of the received view of theory reduction, we can say that a Theory A is said to be reduced by Theory B only when A may be deductively derived from B (Nagel 1961). Often, however, there are terms in A that have no (obvious) corresponding terms in B. Under these circumstances, statements ("bridge laws") are supplied which specify the relationship of terms in A to terms in B. The received view of theory reduction then states that Theory A is reduced by Theory B only when A can be deductively derived from the conjunction of B and the bridge laws. Reduction by deductive derivation is a special case of deductive nomological explanation (Hempel 1965) in which a theory, rather than an event, is the conclusion of the derivation and another, more powerful theory and the bridge laws are the premises (see Nagel 1961).

This account of theory reduction in science has, until recently, enjoyed widespread acceptance within the philosophical community; disputes concerning its appropriateness have for the most part revolved around the problem of how the bridge laws are to be construed. Ager, Aronson, and Weingard (1974:119) point out:

> The necessity of bridge laws has been largely taken for granted in discussions of reduction, and controversy over them has focused on the choice between the three alternative analyses Nagel originally offered: bridge laws are either logical, conventional, or factual connections.

Several philosophers have criticized the received view of theory reduction on philosophical grounds, and some have proposed alternatives (one of which will be discussed shortly). And, Wimsatt (1980b, 1979) has argued that this kind of theory reduction can be found only in principle, *not* in practice. Rather than present the philosophical objections here, we will demonstrate the inadequacy of the view on empirical grounds.

The conception of theory reduction by deductive derivation has some consequences that are unfortunate in the light of current understanding of cognition. The view says that if we have reduced Theory A by Theory B (e.g., reduced classical mechanics by relativity theory), then we understand why Theory A produces usably accurate predictions for a limited set of phenomena; yet at the same time, we understand that Theory A is wrong and unnecessary. Often taken as a corollary of this view is the notion that the reduction of one theory (say, social theory) by another theory (say, neurophysiological theory) renders the concerns and formulations of the reduced theory superfluous. Why study social phenomena as suggested by the weaker theory when this second, more powerful theory explains those things and more besides?

The Structural Basis of Theory Reduction

Sciencing must be viewed as an extension of the basic processes of cognition and examined in the light of our understanding of this process. Several features of these processes are particularly germane here. Cognitive systems develop by passage through a series of invariant stages of complexity. The organism's conceptual structure develops by bracketing—by allowing the organism to make finer and finer distinctions among sensory stimuli, thereby forming and reforming the conceptual schemata that are used for the evaluation of objects or events abstracted from experience. This process of differentiation and integration results in the formation of coherent structures that define a developmental stage. It is important that these stages are ordered hierarchically in development, so that earlier stages are necessary but not sufficient conditions for later ones. Furthermore, structures constructed in earlier stages are incorporated into later ones by transformations and elaborated by empirical modification (see Chapters 1–3).

Theories are schemata that are used for evaluating various stimuli in the environment. Moreover, the analysis of the development of scientific theories shows that they may be viewed as coherent structures produced by the same processes that underlie other aspects of cognition (Piaget 1969). That scientific reasoning characteristically exhibits the same logical structures as nonscientific thought only serves to reinforce this view (Horton 1967; Laughlin and d'Aquili 1974). We suggest that for both the diachronic, developmental progression from one theory to another and the synchronic reduction of theories, the features of the ontogenesis of cognitive structures that we have outlined are strictly preserved. Thus a cognitive process view of theory reduction must make explicit that the reduced theory is necessary to the reducing theory. This means that the move from one theory to the next involves the transformation, incorporation, and elaboration of the reduced theory into the new framework of the reducing theory; unquestionably, the meanings of some terms will be in part transformed (see below) by the new structural matrix in which they are embedded.

Following this line of reasoning, theory reduction in science is analogous to the transformation from one developmental stage to the next in general cognition. It follows the specific course of development taken by the transformation and elaboration of conceptual schemata at any particular stage and the subsequent inclusion of these schemata in higher levels of structural development. This view is fundamental both to conceptual theories of development (e.g., those outlined in Chapter 3) and to the view of the relationship between conceptual structures and sciencing sketched in Chapter 1.

In the received view of theory reduction, the reduced theory may be considered unnecessary because we can explain all the phenomena for which it accounts by reference to the reducing theory. For many it follows, either explicitly or implicitly, that where interdisciplinary theory reduction is concerned, not only the reduced theory itself but also the entire set of concerns of the field of study whose theory has been reduced becomes superfluous (see Wilson 1975). It is perhaps this widespread conception of theory reduction more than any other factor which has led behavioral psychologists to reject attempts by cognitive psychologists to deal with behavioral phenomena (Skinner 1969), social anthropologists to rally periodically against "psychologisms," and Durkheim, Lévi-Strauss, and others to look askance at possible biological bases of structure. To the detriment of all, exchange of theory, insight, and observation has been obfuscated.

Because it requires recourse to the process of equilibration between systems internal to the organism and the environment, the cognitive process view advanced here disallows any arbitrary constraints upon empirical epistemology. It would reject both the antireductionist position that sociological or psychological "facts" require sociological or psychological explanations respectively (Durkheim 1950) and the opposite, orthodox logical positivist position requiring science to reduce to "ultimate constituents," a view associated with the logical atomism of Russell (1956) and the early Wittgenstein (1961). The former position is based upon the naive conception of the nature of systems, summarized by the now classic cliché "The whole is more than the sum of its parts." The latter is based upon the naive view that a sufficiently diligent inquiry will divulge the concrete elements from which the higher levels of system are formed.

With reference to the first position, which might be called the "sociological fallacy," we agree with Buckley (1967) and Blalock (1969) that insofar as the parts and the relations between the parts of a system have been explicated, one has defined the whole. It is, in other words, the job of science to "reduce" to the parts *and* relations comprising a system. As Hinde (1970:7) has said:

> The material with which we start is usually at the behavioral level. If the prediction of behavior, given the antecedent conditions, was the whole aim, there might be no need to reduce to a psysiological level: reference to underlying mechanisms would be unnecessary. But even if the complete prediction of behavior were possible, we should still have advanced only one stage towards its full understanding: a further stage would be reached if the regularities in behavior could be understood in terms of the psysiological organization which they reflect. Thus hypotheses must be judged not only at the behavioral level, but also in terms of the compatibility with lower ones [and higher ones].

To the logical atomist position there can be no better response than that of Whitehead (1960): that if the universe is constructed of organismic processes in an infinite concatenation of systems within systems, the search for a level of ultimate constituent elements is futile. The level or levels of systemic organization one chooses for analysis is therefore not a matter of social norm but rather *is logically entailed by the problem one has set.*

The life sciences are concerned with the nature of organic systems, systems that are comprised of organized subsystems whose arrangements emerge in equilibration with their environments. From this and other considerations mentioned above we may stipulate what we will call the *rule of minimal inclusion: Any explanation of behavior must take into account any and all levels of systemic organization efficiently present* ("statistically relevant," in Salmon's [1971] sense; see Chapter 6) *in the interaction between the system operating and the environment of that system.* The rule of minimal inclusion will require the theoretical consideration of systemic levels at least one step below and one step above the level or levels appropriate to the phenomenon being explained (compare with Wimsatt's [1980a, 1980b] arguments for multilevel explanations). Rather than requiring the obsolescence of reduced theories, the rule of minimal inclusion requires the structural merger of reduced and reducing theories when they account respectively for different levels of systemic organization of interest for the explanation of the behavior of interest. Thus, for example, the reduction of cognitive theory by neurophysiological theory would require the incorporation of both into a single theoretical structure. To reduce one theory by another, then, is to cut across theories (and often disciplines defined by those theories) so that the conceptually relevant notions in one theory are unified with the conceptually relevant notions in the other. It thereby becomes possible to speak of conceptual schemata and neurophysiological models ("engrams," "cell assemblies," and so on) in the same breath without threatening the importance of either theoretical approach. As a matter of fact, this view of theory reduction forces the scientist to focus attention more sharply on the several theoretical approaches impinging upon the problem at hand (for applications of this view see d'Aquili et al. 1979; Hufford 1982; Rubinstein 1976, 1979b, 1984).

This view of what reduction ought to be like can be called a structural elaboration model. The implications of this model are many. Principally it appears that development through structural elaboration is a process of both quantitative and qualitative change. Thus, for example, although there are quantitative changes between Stage 3 and Stage 4 structures (see Figures 3.1 and 3.2 in Chapter 3) that may be measured on an additive scale, the increased interconnectedness of the structure has profound qualitative ramifications for the stability and flexibility of the structure and for the mode of information-processing it employs. This kind of qualitative shift, due to quantitative growth, has been demonstrated in both biological evolution (Rensch 1959; Laughlin and d'Aquili 1974) and developmental semantics (Rubinstein 1976), and its operation in the transition between states of consciousness has been suggested (Tart 1975:243–57). A structural-elaboration view of reduction suggests that the combinatory principles that function during the incorporation of theories from different levels are distinct from the additive principle underlying the more traditional view of combining theories such that the joining of two theories each with a probability of .70 yields a less satisfactory theory with a probability of .49 (Martindale 1977). Thus an *a priori* rejection of reduction as we view it, based on the fear that the "correctness" of the multilevel theory will be less than that of its component theories, is unwarranted.

Viewed as a process of structural elaboration, the reduction of theories requires the structural merger of the reduced theories. A by-product of the resulting change in the structural matrix in which the theories are placed is the conceptual revision of the reduced theories.

Applied to anthropology in general, this view of theory reduction provides two cogent points: (1) During the analysis of an anthropological problem, it is not necessary that more "traditional" anthropological concerns be abandoned. Rather, the relevant data are to be distilled from the traditional approaches, focused, and combined with data from less traditional approaches in the effort to resolve the problem at hand. (2) The data, insights, and theories from all levels of systemic organization contributing to the resolution of the problem at hand may and must be incorporated into a single, coherent analysis.

Hanson (1958), Kuhn (1970, 1974), Feyerabend (1965), and others have argued that scientists working within a particular theoretical framework have no way of understanding the terms and phenomena of other frameworks. Hanson attributes this to the development of a gestalt, a way of viewing the world that prevents the scientist from seeing the world in the same manner as his or her cross-theoretic colleague. He calls this phenomenon to our attention to emphasize the fallaciousness of the distinction between the context of discovery and the context of justification—he offers the distinction between "seeing" and "seeing that." The thrust of Kuhn's and Feyerabend's treatments of this problem has been to attack the "layer-cake" conception of the development of science. If a term (e.g., culture) has different meanings in Theory A (e.g., Tylor's evolutionism) and Theory B (e.g., Geertz's hermeneutics), in what sense can the latter be said to build on the former? Here we will consider the problem only briefly in order to indicate how it may be seen from our view of theory reduction.

Relevant conceptual material from the reduced theory is incorporated into the reducing theory. Unquestionably, not all of the content of theoretical terms will be shared, partly because of differences in the structural matrices in which they are found and partly because of differences in the problems being addressed. Neither will their contents be completely incommensurate, however, and while the exact degree of meaning overlap in a specific case is a matter for empirical and logical demonstration, we suggest that it is often great. The reason is a structural one. Scientific concepts derive their meaning from their structured, conceptual context of the theory in which they are embedded. If two theories have the relationship to each other of reduced and reducing theories, then the reduced theory is structurally incorporated into the reducing theory, thus providing a single theoretical framework for relating the meanings of terms in the two theories. Part of what we suggest, then, is that the contextualists have presented a false dilemma because, in large measure, their accounts of the intertheoretic incommensurability of terms lack a biopsychological underpinning. The choice is not between complete and simple absorption of theoretical terms from one theory to another (which leaves, as the contextualists have certainly demonstrated, an unrealistic account of scientific progress) and total incommensurability of meaning of theoretical terms (leading to inevitably narrow gestalts). Nor, as some people have suggested, is there a compromise to be made via the logical structure of theories alone leading to partial commensurability of terms. Rather, we suggest, at least partial commensurability obtains because of the nature of the biopsychological processes mediating scientific inquiry.

Reduction by Incorporation

Having found the received view of theory reduction wanting for a variety of biopsychological reasons, and having outlined a few of the biopsychological considerations important to developing a more sophisticated account of theory reduction in science, we would like to summarize a philosophical approach to some of the problems raised.

Ager, Aronson, and Weingard (1974) and Aronson (1984, n.d.) have outlined an approach to theory reduction that they have termed "reduction by incorporation." Starting from the premise that "to be scientifically respectable, bridge laws should be more exposed to experimental test" (Ager et al. 1974:121), they demonstrate that the received account allows the adequacy of bridge laws to be decided on logical rather than empirical grounds. They propose that reduction can occur through the incorporation of diverse scientific systems by means of reducing theories. To do this they modify the traditional account of reduction by replacing bridge laws with identity statements of the form (for the reduction of mental [M] and neurophysiological [N] events) (P) [MP ⊃ (P = N_1 v P = N_2 v . . . v P = N_n)]. In the traditional view of reduction, the identity statements are premises. In this view, they are theorems which, along with the laws of the scientific systems being reduced, tell us the possibilities of identity. Leibniz's law allows the identities to be pinpointed.

Reduction by incorporation can be clarified by considering the specific case of reduction involving neurophysiological and mental material presented by Aronson (1984, n.d.; see Ager et al. 1974 for a treatment of the reduction of empirical gas law and kinetic theory). Aronson proceeds: "(II) (P) [MP ⊃ (P = N_1 v P = N_2 v . . . v P = N_n)]; (1) laws relating to neurophysiological states; (2) laws relating to mental states; (3) identity(ies) between a mental state and a neurophysiological state." He continues: "If we discover for example that there are neurological and mental states that are related to other states in exactly the same way—say, $N_{25} = f(N_3,N_4)$ and $M_{10} = f(N_3,N_4)$ and $f = f$ then we can eliminate various disjuncts in the consequent of (II) and pinpoint our identities, in this case (3) $N_{25} = M_{10}$."

Aronson notes that the use of identities in reduction has certain desirable consequences not provided by the use of correlation: "It becomes clear . . . that identities can explain things that correlations cannot because the former has implications which the latter lacks, viz.; (1) $M_{10} = N_{25} \supset (x) (M_{10}x \equiv N_{25}x)$ while (2) $M_{10} \equiv N_{25} \not\supset (x) (M_{10}x \equiv N_{25}x)$. Thus correlation cannot explain in any non-*ad hoc* manner why M_{10} and N_{25} are states of the same individual or are located at the same time and place whereas $M_{10} = N_{25}$ renders such explanations elementary."

An important ramification of the reduction-by-incorporation view is that it allows the incorporation of information from diverse scientific systems in finding the solution to a problem. A corollary ramification is that, unlike correlational approaches, this one does not support a dualistic ontology—that is, does not encourage what Whitehead called the "fallacy of the bifurcation of nature."

Since this approach to theory reduction allows the elaboration, transformation, and incorporation of concepts across theories, it is consonant with our present understanding of the development of conceptual systems in ontogeny. It is therefore conducive to a more sophisticated, cognitive process view of scientific theory reduction.

All knowledge is incomplete—fallible in Peirce's sense (Almeder 1973, 1975). Knowledge may be fallible because of incomplete information content (the most common sense of the term) or because the structural organization of that information is different than the reality of concern. Either way, with Pribram (1971) we would pay utmost attention to anomalies and "paradoxes." Reduction by incorporation is not an end in itself. Rather, it is a means to the most complex model-building available to us at the moment, one that has a reasonable chance of transcending the more deleterious effects of disciplinary involution. Further, recognition of epistemological fallibility requires that we constantly maintain conscious distinctions between (1) operational logic (how the cognitive system processes information about the organism and its world) and cognized logic (the fallible but conscious model of how the cognitive system processes information); (2) "content" (information about the world) and "structure" (the organization imposed information by the cognitive system); and (3) the "cognized environment" and the "operational environment."

This chapter treated the cognitive process features underlying sciencing outlined in earlier chapters as a minimal set of primitives with which an adequate view of the nature of theory reduction must agree. The traditional view of theory reduction is not compatible with these features. Further, it appears that a view of reduction compatible with them would eliminate several pernicious results of the traditional view. In particular, the reduction of one theory to another does not require the obsolescence of the reduced theory. Also, data and theories from different levels of systemic organization need to be structurally unified in order to facilitate the solution of problems in the life sciences. Structural unification of theories, including reducing and reduced theories, results in partial commensurability of meaning between terms of different theories. An extant philosophical approach to theory reduction has been outlined and presented as a possible logical infrastructure for a more biopsychologically sophisticated account of theory reduction.

Theory reduction is often taken to be a special kind of explanation in science. As such the view of reduction presented in this chapter presages the conception of explanation that follows from the cognitive process considerations outlined earlier. It is to a more detailed look at explanation as cognitive process that Chapter 6 turns.

Six

Explanation and Cognition

This chapter explores more fully what a model of explanation consistent with the view of cognitive process described earlier might look like. A review of the nature of explanation in anthropology is taken as an exemplar for this task. This chapter first shows how a number of models of explanation are explicitly and implicitly accepted in anthropology. Then it summarizes the critical philosophical debate surrounding these models and suggests that the currently prescribed models are inadequate for anthropological purposes. The constraints on scientific explanation that derive from the model of scientific cognition are briefly reviewed. An account of explanation which emerges from the philosophical debate, and which is consistent with the cognitive process view of sciencing, is sketched. Finally, we show that this account is a good descriptive account of explanation in anthropology as well as a good prescriptive model of scientific explanation.

Anthropology and Explanation

It is useful to start by considering the "received" view of scientific explanation—the Deductive-Nomological (D-N) model of explanation, and its modification, the Inductive-Statistical (I-S) model of explanation—and its distribution in anthropological work (see Hempel 1965; Nagel 1961; Suppe 1977; cf. Fritz and Plog 1970; Spaulding 1968; Jarvie 1967; Watson et al. 1971).

The D-N and I-S approaches to explanation have had their fullest development in the work of Carl Hempel (1962, 1965) and for that reason are referred to as the Hempelian models of explanation. The paradigm case of this form of explanation is the D-N model, which can be characterized as follows: an investigator poses a question that specifies the phenomenon to be explained (the *explanandum* event). The explanation of that phenomenon takes the form of an argument consisting of the explanandum and a number of true premises (the *explanans*) that deductively entail it (that is, where if the premises are true, the conclusion must be true by virtue of logical form). Further, among the premises from which the explanandum is derived there must be at least one universal law which is essentially contained, and at least one empirically observable fact (or "initial condition").

The following, for example, is a D-N explanation:

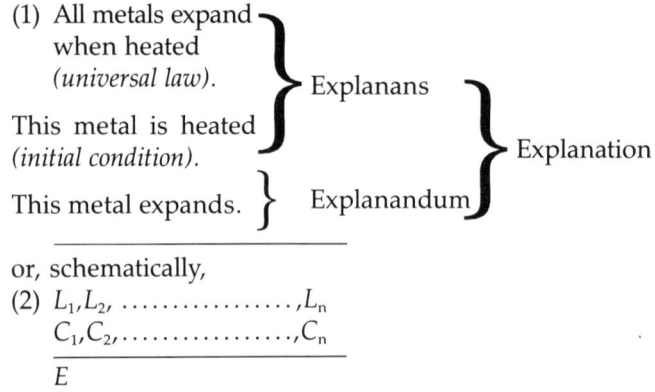

(1) All metals expand when heated *(universal law).* } Explanans

This metal is heated *(initial condition).*

} Explanation

This metal expands. } Explanandum

or, schematically,

(2) $L_1, L_2, \ldots\ldots\ldots\ldots\ldots, L_n$
$C_1, C_2, \ldots\ldots\ldots\ldots\ldots, C_n$
―――――――――――
E

where L is a universal law, C is an initial condition, and E is the explanandum.

Since not all phenomena are amenable to explanation by deduction (Nagel 1961:503–20), Hempel developed a number of forms of statistical explanation based on the D-N model (Hempel 1962, 1965). In developing these models, Hempel proceeded by reworking some of the formal conditions of adequacy of the D-N model. His most important model of inductive explanation is the Inductive-Statistical (I-S) account of scientific explanation.

The I-S model has the same basic form as the D-N model of explanation, but it differs in that

(1) The logical connection between the explanans and the explanandum is inductive, rather than deductive.
(2) A statistical generalization, rather than a universal law, must be essentially contained in the explanans.
(3) The explanandum should follow from the explanans with a moderately high probability.

Thus an example of I-S explanation might be:

(3) There is a moderately high probability that bacteria which come in contact with penicillin die.
$$\frac{\text{This bacteria has come in contact with penicillin.}}{\text{This bacteria is dead.}} = r$$

or

(4) $$\frac{SG_1, SG_2, \ldots\ldots\ldots, SG_n}{E} \frac{C_1, C_2, \ldots\ldots\ldots, C_n = r}{} = r$$

where SG is a statistical generalization, C is an initial condition, E is the explanandum, and r is some level of probability.

Two features of these models of explanation need to be emphasized, since they will figure in the later critical discussion of the D-N and I-S models. First, both construe explanation as argument. Thus, for the D-N model, an explanation is an argument which shows that the event to be explained was necessary, on the basis of particular facts and general laws, due to the deductive connection between these facts and laws and the event. In the same way, for the I-S model explanations are construed as arguments which show that the event was *likely*, given the particular facts and statistical generalizations, due to the strong inductive link between them and the event.

With this understanding of the D-N and I-S models of explanation, we turn to some examples of how the use of these models of explanation has been advocated, both implicitly and explicitly, in anthropology. First, consider two examples of the explicit normative application of these "covering law" models of explanation to anthropological problems.

Prompted by a number of sociological considerations, the "new archaeology" was accompanied by a dramatic self-consciousness about the status of "laws," "theories," "processes," "explanation," and the like in archaeology, and about what that status should properly be (Plog 1973: 649–51). This self-consciousness brought a wave of reflection on the issues that an examination of these concepts entails (e.g., Binford 1965; Binford and Binford 1968). In a sense, this philosophical interest reached its apex with the publication by John Fritz and Fred Plog in 1970 of an article in which they claimed that the D-N model of explanation is the appropriate normative model for archaeological inquiry. They said:

> We argue that the Hempel-Oppenheim model of scientific explanation is, at worst, an important heuristic device which provides insight into the structure of archaeological knowledge. At best it points the way archaeologists must travel if they are to contribute to the corpus of laws of human behavior. (Fritz and Plog 1970:406)

While the proposals that Fritz and Plog advanced are not unanimously embraced by archaeologists, the objections that have been raised to their proposals revolve around issues of the difficulty of practical application (Chenall 1971), notational form (Tuggle et al. 1972), or of the relationship of the philosophy of science to science (Levin 1973), rather than around substantive criticisms of the normative appropriateness of the D-N model per se.

A second, equally explicit call for the use of D-N models of explanation in anthropology has been voiced by I. C. Jarvie (1967). In *The Revolution in Anthropology*, Jarvie advances as a satisfactory resolution of the "problem of cargo cults" an explanatory account based on Popper's "rationality principle." It includes as an essential premise a law of the sort: "Agents always act appropriately to their situations as they perceive them." As initial conditions it includes (1) a description of the situation—Agent X perceived himself to be in situation S—and (2) an analysis of the situation—the appropriate action for any agent in S is A (Koertge 1975).

Although Jarvie calls his version of the rationality principle by the name "situational logic," it is clear from his programmatic comments and his analysis of cargo cults (1967:106ff.) that he has adopted Popper's rationality principle with little modification. Thus, early in his analysis he notes, "An explanation . . . consists of deriving a statement describing what is to be explained from another statement" (1967:17). And later in his discussion he says:

> Popper argues that our idea of explanation can be given fairly precise logical meaning which amounts to deriving the fact to be explained from a universal law or laws, combined with a set of factual statements or "initial conditions." (1967:79–80)

Jarvie clearly advocates adopting a form of the D-N model of explanation for anthropological use.

We turn now to slightly more opaque instances of the application of the D-N model of explanation to anthropological problems. Schneider's (1974) statement of formal economic anthropology may be taken as an example. Schneider endeavors to construct a rationale for the anthropological use of the formal models of microeconomics, and for the "deductive method" in general. While we think that Schneider harbors several misunderstandings about prediction, the role of induction and deduction, and the nature of theories (see

Chapter 4), his treatment of formal economic anthropology is nonetheless instructive for the present discussion.

Throughout *Economic Man*, Schneider urges that the explanation of (economic) behavior should take roughly the following form. To explain a particular phenomenon or set of phenomena, an anthropologist should identify a number of particular variables in the social situation being studied (these must meet certain requirements; e.g., quantifiability). These variables, combined with the "laws" of microeconomics, then allow the deductive derivation of the statement of the phenomenon in question. The concepts of microeconomic theory play the role of general laws in Schneider's formulation, while the variables that are identified serve as the initial conditions. Further, the conjunction of these two sorts of information allows the deduction of an *explanandum* event. This is a special case of the D-N model of explanation, wherein the role of the general laws is always played by propositions or generalizations drawn from microeconomics. For Schneider, an adequate explanation of economic phenomena has the following form:

(5) General law(s) from microeconomics.   } Explanans
Initial conditions (parameters imposed by particular society).
———
Phenomenon to be explained. } Explanandum

We think it clear that the covering law models of explanation presented by Hempel have been widely proscribed in anthropology. At the risk of belaboring this point, and because it is particularly relevant to what we said in Chapter 1, we here briefly consider the use of the I-S model in cognitive anthropology.

The quickest way to see the use of the I-S model of explanation in cognitive anthropology is to look at two aspects of that approach. First, cognitive anthropology makes the assumption that human cognition and behavior is rule-following in the sense that behavior is generated by the application of a set of rules or principles to specific situations (see Chapter 1).

Second, for cognitive anthropology, human cognition and behavior are considered explained when we can array what are supposed as the informants' rules and detail the circumstances under which these rules are used. Taken together, these allow the derivation of the ethnographer's corpus of data.

Thus an informant's use of kinship terminology would schematically be explained like this:

(6)   *SG1*. In the informant's culture, the rules of use of kinship terms are *X*.

    SG2. In the informant's society, when conditions *Y* are present the speaker will use kinship terms according to the rule more often than not.

    C1. <u>Conditions *Y* prevail.</u>

     E. The informant used the kinship terminology in this (an appropriate) way.

Again the explanation is presented as ideally being an argument that tells us that the explanandum event was to be expected given the explanans.

The following section reviews briefly some of the objections that have been raised to these models of explanation, and it indicates why they ought to be considered inadequate for anthropological purposes.

Critiques of D-N and I-S Explanation

The Hempelian covering law models of explanation derive directly from the "received view of scientific theories" (see Suppe 1977). In general, the received view holds that scientific theories are to be construed as "axiomatic calculi which are given partial observational interpretation by means of correspondence rules" (Suppe 1977:3; cf. Braithwaite 1953). This view of the nature of scientific theories is in the Logical Positivist tradition of philosophical analysis and, in particular, emerged from the work of the Vienna Circle and the Berlin School. Now mainly fallen into disrepute, this view (although it underwent some internal repair) continued to be the accepted philosophical analysis of scientific theories for some time after the major tradition from which it had developed had declined. As Suppe (1977:3–4) points out, it was not until the 1950s that

> this analysis began to be subject to critical attacks challenging its very conception of theories and scientific knowledge. These attacks, which continued well into the 1960's, were of two sorts: First, there were attacks on specific features of the Received View . . . designed to show they were defective beyond repair. Second, there were alternative philosophies of science advanced which rejected the Received View out of hand, and proceeded to argue for some other conception of theories and scientific knowledge. These attacks were so successful that by the late 1960's a general consensus had been reached among philosophers of science that the Received View was inadequate as an analysis of scientific theories; derivatively, the analyses of other aspects of the scientific enterprises (for example, the notion of partial interpretation, [explanation] and the observational theoretical distinction) erected upon the Received View became suspect and today are subject to much the same skeptical criticism.

The criticisms of the D-N and I-S models of explanation that are summarized here, and the alternative account of explanation to be presented later, are part of this skeptical examination of the received account of the acquisition of scientific knowledge.

The D-N and I-S models of explanation construe explanation as arguments that show that the explanandum event was to be expected. While this characterization of scientific explanation is not fully satisfactory when the inference is made on the basis of deductive logic, it is especially problematic for explanation which bases the inference on inductive logic, as does the I-S model. The reason this is so is twofold; the Hempelian covering law models require that explanations be arguments. However, as Jeffery (1969:22) points out, the fact that we can sometimes explain phenomena (i.e., gain *knowledge why*) by providing inferences that give "knowledge that" is not a strong reason to require that *all* explanations be inferences. Further, not all phenomena are amenable to explanation via inference. This is particularly so for the explanation of improbable events like the one reported in the following news item (cited in Ornstein 1978:60).

> FISH KILLS SEAGULL
> *Brixham, England* (UPI). Members of a yacht club say they saw a fish kill a seagull. They said when the bird dived Wednesday to grab a fish in Brixham Harbor, the fish grabbed the seagull, pulled it beneath the water and drowned it.

In cases like this, Jeffrey (1969:25) notes:

> Inference is *not* an explanation . . . by the fact that even when the improbable chances to happen, we give the same sort of account: the happening was the product of a *stochastic* process following such-and-such a probabilistic law. And, we gloss this by pointing out that in this case the unexpected happened. My point is that it is no less a gloss, and no more essentially part of the explanation, when we point out in the more usual cases that the expected happened.

A further criticism is that although there is no adequate account of how inductive inferences are to be made, practicing scientists produce workable explanations quite routinely. Salmon (1971:37) points out, "There is no analogue of the [deductive] rule of detachment *(modus ponens)* [in inductive logic], so inductive logic does not provide any basis for asserting inductive conclusions." This lack of rules of inference, combined with other features of inductive logic, create several problems for the I-S model of explanation. Particularly, Salmon (1971:37) points out, inductive logic provides no means for producing statistical generalizations as detachable conclusions of any argument; inductive inferences can proceed from particular to particular, so there is no reason to require inductive explanations to include general statement in the "premise"; and inductive inferences express (it appears) statements of degree of confirmation rather than the usual relation between explanandum and explanans in an argument. It also appears mistaken, then, for I-S explanation to require that the explanatory enterprise take the form of argument because we cannot adequately characterize what is entailed by the notion of "inductive inference." The first aspect of the Hempelian approach to explanation that is recalcitrant to philosophical analysis is the notion that explanations must only be explanation by argument.

The second problem with the D-N and I-S models of explanation is their requirement that the explanans contain general or statistical laws. While it may seem relatively unproblematic to apply this requirement to statistical explanation, it is highly problematic when applied to D-N explanation and is subject to criticism on two levels. The first is that, even if one accepts Hempel's formal grounds, explanations can be given without the use of laws. Aronson (1969, 1984), among others, provides one such critical account.

Moreover, an adequate characterization of what it is to say that X is a universal law has yet to be offered. Without such a characterization, we are left uncertain as to how to interpret the D-N model's insistence that at least one universal law appear in the explanans.

Yet another aspect of the I-S model which may be subject to criticism is the requirement that an inductive explanation show that the explanandum was to be expected on the basis of the information provided in the explanans, with a moderately high probability. The problem here is this: What justification can be offered for choosing some particular level of probability as the criterion?

The problem with this requirement, as Scriven and others have noted, is that there are phenomena (like the news item reported earlier, or the patterns of exchange characteristic of the Ik family [Turnbull 1978]) that have a low probability of occurring which, presumably, we might want to be able to say we can explain. This point is particularly important for explanation in anthropology, because anthropologists are likely to continue to find that some behavior, aspect of behavior, or record of behavioral activity, to the occurrence of which they had ascribed a low probability, in fact occurs. Presumably, our ethnographic accounts and analyses would not be considered complete if we were unable to produce "legitimate" explanations of these rare phenomena. This is all the more important because the examination of these improbable events may yield valuable information about human behavior (Laughlin and Brady 1978; Rubinstein and Tax 1981).

The Hempelian account of explanation consists of providing a (formal) argument in which the conclusion is likely to (or necessarily does) follow from the premises. On this account we would need to say that the more likely the explanans makes the explanandum event appear the stronger the explanation. But this need not always be the case. Consider Greeno's (1970:89) example (we have rewritten Greeno's probability notation so that it will agree with the notation used in the latter part of this chapter):

> Suppose that a boy, Albert, is convicted for stealing a car. Attempting to give an explanation, a social worker points out that Albert lives in San Francisco, where there is a high delinquency rate. However, it is also noted that Albert's father earns $40,000 per year, and sons of men with high incomes have a low delinquency rate.
>
> One view which agrees with intuition and most recent discussions of statistical explanation [i.e.,

Hempel's account of I-S explanation], is that a "good" statistical explanation allows us to assign a high probability or likelihood to the event to be explained. Let S be satisfied by a young man living in San Francisco, and M be satisfied if a young man is convicted of a major crime. Then we have the statistical hypothesis,

$$P(S,M) = p$$

where p is some rather high probability. And we carry out a simple argument of the form

(1) $P(S,M) = p$
$\underline{a \in S}$
$L(a \in M) = p$

where $L(a \in M)$ stands for the likelihood that individual a is in the set (has the property) M. Since we take it that p is fairly high, argument (1) should be taken as a fairly good statistical explanation of Albert's delinquency.

The difficulty with this is that we have not included an item of relevant information, the income-level of Albert's father. Let H be satisfied if a boy comes from a family with more than, say, $30,000 per year. Then we have,

$$P(S \cdot H, M) = p'$$

where p' is considerably lower than p (and probably lower than $P(M)$ in the whole population). And the argument becomes

(2) $P(S \cdot H, M) = p'$
$\underline{a \in S \cdot H}$
$L(a \in M) = p'$

The difficulty here is clear. If, with Hempel, we require that statistical explanation makes the explanandum seem probable, then Greeno's (1) is clearly a stronger explanation than his (2). Yet, if we are interested in increasing our understanding of the explanandum event's occurrence (interested in gaining "knowledge why"), then Greeno's (2) is a better explanation than his (1).

The issue is this: Should explanation be construed as a process of information increase (or reduction of uncertainty), or should it be construed as an exercise in the construction of arguments (on the basis of which we might expect the phenomenon to occur)? While the latter conception of explanation may be a useful way to look at some class of explanations, in our view the majority of explanations should be seen as statements that increase information, or knowledge, about phenomena.

The Problem of Explanation in a Cognitive Process Approach to Science

In addition to any reservations about the appropriateness of the D-N and I-S models of scientific explanation that arise from the preceding discussion, there are empirical considerations that lead to the conclusion that these models are inadequate normative or descriptive accounts of explanation in science. This section briefly reviews the cognitive process view of science presented earlier, from which these empirical considerations come. In the course of this review, the empirical inadequacies of the D-N and I-S models are revealed, and an understanding of the process of sciencing with which a theory of explanation must be consistent emerges.

Although faced with a chaotic array of stimuli, in supplying order to its experience of the world Homo sapiens, like all higher organisms, makes use of and attends to only a small range of the variety present. Order is imposed on the variety when sensory input is integrated into an individual's cognized environment. These models are subject to ongoing modification as the individual interacts with the operational environment roughly as follows. Based on the information in his or her models, an individual generates expectations about the results of his or her interactions with the environment. The resulting behavior tests these expectations, and information about their success or failure is derived. Depending upon the degree of discrepancy between the expectation and the results, the models are modified to come into more adaptive isomorphism with the environment (see Chapter 1 for a discussion of adaptive isomorphism). Sciencing is the extension of these fundamental processes.

Because of this relation it is important to examine explanation in sciencing in light of our understanding of these fundamental biopsychological processes. Of these processes, those related to ontogenetic development are particularly informative (see Chapters 2–4).

On the view of sciencing employed here, theories are schemata used to evaluate various stimuli in the environment. Moreover, the formation and refinement of theories is seen as analogous to the developmental processes by which an individual generates successive models of the environment. It is because the growth of these models involves the development of a means for making finer and finer partitionings of the environment that they serve to enhance the adaptive isomorphism of an individual's models with the environment. Further, the articulation of these models, and the progressive partitionings of the environment, result in changes in the individual's understandings and accounts of the environment, physical, social, and symbolic.

When we attempt to account for explanation in sciencing, it is crucial that we preserve at least two aspects of the processes outlined above. Our account of explanation must provide for the growth of understanding via progressive bracketing, and it must capture the alteration of understanding (the change in "knowledge why") that follows from this process.

The ontogenetic growth of an organism occurs on many levels of organization—biological, cognitive, affective, symbolic, social, and so on—simultaneously. As a result of this, behavior derives from the interaction of structures at several levels of organization, and an individual's understanding of the environment necessarily reflects this multivariable development. As noted in Chapter 5, this results in the need for any adequate account of behavior to consider any and all levels of systemic organization efficiently present in the interaction between the organism behaving and the environment of that organism. Thus any full (normative and descriptive) account of explanation in sciencing must at least allow, and at best constrain, explanations to be multilevel in this sense; therefore facilitating the construction of complex models of a complex world (Wimsatt 1981).

Both because of their form and logic, the D-N and I-S models of explanation fail to meet the minimal empirical requirements set out above. In the following section, we describe an alternative model of explanation available in the philosophical literature which meets these requirements, overcomes the earlier outlined philosophical difficulties, and thus appears to provide a useful logical infrastructure for the understanding of explanation in sciencing.

Statistical-Relevance Explanation

A model of explanation consistent with the concerns discussed in the preceding section is the Statistical-Relevance (S-R) account of explanation developed by Wesley Salmon (1971). This account views explanation as the reduction of uncertainty, or the increase in information about the phenomenon to be explained.

In the S-R account, explanation begins with the posing of an explanatory question of the form: Why does this x which is a member of A also have the property B? (Salmon 1971:76). Rather than expecting the answer to be an argument that makes the explanandum event seem probable, the explanation is informally characterized as a process that involves taking the class of things to which the phenomenon to be explained belongs and dividing it into a set of disjoint subclasses. The probability that the members of each of these subclasses has another property X is then determined, and the phenomenon being explained is referred to one of the subclasses. Formally, S-R explanation may be characterized as a partition of the reference class A into a number of subclasses, all of which are homogeneous with respect to B, along with the probabilities of B within each of these subclasses. In addition we must say which of the members of the partition contains our x. More formally, an explanation of the fact that x, a member of A, is also a member of B would go as follows:

$$P(A \cdot C_1, B) = p_1$$
$$P(A \cdot C_2, B) = p_2$$
$$\cdot$$
$$\cdot$$
$$\cdot$$
$$P(A \cdot C_n, B) = p_n$$

where

$A \cdot C_1, A \cdot C_2, \ldots A \cdot C_n$ is a homogeneous partition of A with respect to B

$p_i = p_j$ only if $i = j$, and $x \in A \cdot C_k$

(Salmon 1971:76–77)

For purposes of this discussion, four aspects of this characterization of explanation should be noted. First, a probability statement is a two-place function which states the probability that a thing x, which is a member of a certain class of things A, is also a member of another class B. The first class A is the reference class, while the second class B is the attribute class. In Salmon's notation:

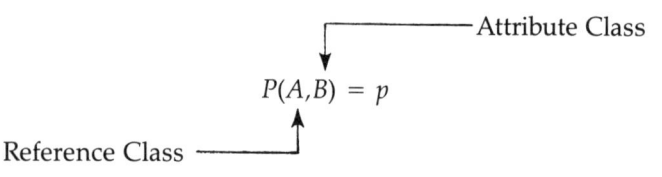

$$P(A, B) = p$$

Attribute Class — (arrow to B)
Reference Class — (arrow to A)

Second, to partition a class, all that need be done is to divide the class into subclasses so that every member of the class belongs to one and only one subclass (see Suppes 1966). Thus for a class A, which contains only the members 1, 2, 3, 4, we can say $\{A\} = \{1, 2, 3, 4\}$. Hence, the class of subclasses $\{A_1\}$, $\{A_2\}$, $\{A_3\}$, $\{A_4\}$ is a partition of A just in case two conditions are met:

1. The subclasses are disjoint, i.e., the intersection of any two of the subclasses of the partition is empty
2. The subclasses are exhaustive; e.g.,

$\{A_1, A_2, A_3, A_4\} = \{A\}$

The third notion is the concept of statistical relevance. "A property C is said to be *statistically relevant* to B within A if and only if $P(A \cdot C, B) \neq P(A \cdot B)$" (Salmon 1971:42). This is to say that a particular property is statistically relevant to the occurrence of the attribute in the reference class just in case it *changes* the probability (either up or down) that the attribute occurs.

Fourth, it is a formal requirement of the S-R model of explanation that the partition of the reference class be homogeneous with respect to the attribute class. This is to say that there is "no way, even in principle, to effect a statistically relevant partition without already knowing which elements have the attribute in question and which do not" (Salmon 1971:43). There are two weakened versions of the homogeneity of the partition of the reference class rule that should be noted. A partition is said to be *epistemically homogeneous* if "we know or suspect that a reference class is not homogeneous, but we do not know how to make any statistically relevant partition" (Salmon 1971:44). And a partition is said to be *practically homogeneous* when "we know what attributes would effect a statistically relevant partition, but it is too much trouble to find out which elements belong to each subclass of the partition" (Salmon 1971:44), or where we have neither the techniques nor the ability to repartition the reference class in the homogeneous fashion.

Thus, S-R explanation requires that when we partition the reference class we do so in a fashion that excludes further statistically relevant partitioning. Since a truly homogeneous reference class is preferable to either an epistemically homogeneous or practically homogeneous partitioning of the reference class, it is incumbent on the investigator to repartition the reference class (i.e., reanalyze the data) when either new knowledge or new techniques make this possible, even if this results in making the events of interest appear less likely.

There is one more formal relation deriving from Salmon's presentation of the S-R model which should be noted. This is the screening-off relation. Salmon (1971:55) provides a formal definition of this relation. He says: "We may say that D screens off C from B in reference class A iff (if and only if) $P(A \cdot C \cdot D, B) = P(A \cdot D, B) \neq P(A \cdot C, B)$." In other words, the presence of attribute D prevents attribute C from influencing the probability of the event in the way it can in D's absence. This relation provides a formal criterion on which to base a choice between alternative explanations. Its utility becomes clear in the following section.

Statistical relevance explanation may be summarized as follows: The investigator poses a question of the form: Why does x, which is a member of the class A, also have the property (attribute) B? By partitioning the reference class into a number of subclasses, and stating which subclass x belongs to, ascertaining the probability that members of each of the subclasses of A will have the property B, the investigator provides an answer in the form: Because x also has the property C and $P(A \cdot C, B) = p$.

This form of explanation encounters none of the problems that D-N or I-S explanation meet. In addition, since explanation is seen as aiming at the increase in information about phenomena rather than as a process of constructing arguments which lead us to expect phenomena to occur, S-R explanation encounters no formal difficulties in explaining improbable events. The

S-R account appears not only to be a good normative infrastructure for our understanding of explanation, but also provides a good descriptive framework for the study of explanation in anthropology.

The S-R Model in Anthropological Research: An Archaeological Case

This section shows that a portion of archaeological research entails the implicit use of the S-R model (this material is from Rubinstein and Donaldson 1975). It is interesting that it is in the work of archaeologists who most actively espouse the use of the D-N model of explanation where the S-R model appears to form a good descriptive account of the research enterprise.

Spaulding (1960:438), among others, claims that "the subject matter of archaeology is artifacts." However, before the archaeologist can say anything meaningful about a collection of artifacts, he or she must sort them into categories appropriate for analysis. Use of the techniques of typology and seriation explicitly require the partitioning of the assemblage collection (a reference class) into (hopefully) homogeneous subclasses. Lithic assemblages, for instance, have been partitioned on the basis of morphology/technology (Bordes 1972) and functional attributes (Binford and Binford 1966); sorting of ceramic collections has been accomplished on the basis of attributes of vessel morphology (Petrie 1899), of technological components (Plog 1976), and of design elements (Deetz 1965); and other approaches have been suggested using functional attributes (Read 1974) or "non-typological" ordering using multidimensional scaling (LeBlanc 1975). These examples may suffice to illustrate that, while the specific aims of particular investigators vary greatly, when they proceed by partitioning the reference class into meaningful (i.e., statistically relevant) subclasses information is substantially increased. Explanation of human behavior based on archaeological data is given by providing meaningful partitionings of the class of artifact assemblages into subclasses which are, ideally, disjoint and exhaustive. It is important to note that the theoretical motivation for the use of artifact typologies is generally accepted within archaeology; debates center on issues of which attribute classes will provide statistically relevant partitions (types) of the reference class (a collection of artifacts).

Turning to a more specific example, consider the research design prepared for the Southwestern Anthropological Research Group (SARG) by Fred Plog and James Hill (1971, 1972). Since the participants in SARG had diverse research interests, the research design and methods were formulated on a level broad enough to embrace the efforts of each investigator. The group's goal was one of "explaining the variability in the spatial distribution of prehistoric sites" (Plog and Hill 1971:7). The basic design sets forth a minimal set of tasks to be undertaken in reaching this goal. These tasks amount to specifications of site location vis-à-vis particular variables in the natural environment, for example, landform, drainage rank and pattern, and plant community. In other words, each investigator's study area was to be subdivided (partitioned) on the basis of a class of natural factors that were felt to be determinants of site location. The reference class was to be partitioned into subclasses on the basis of posited statistically relevant attributes.

Plog (1972) offers a case study in the application of the research design, drawing upon transect-survey data from the Chevelon drainage in eastern Arizona. Plog examines the distribution of two types of sites (all sites and settlements) in relation to two partitionings of the Chevelon drainage (the reference class). His first partition is made on the basis of plant community and the second on the basis of biome in conjunction with proximity of water resources.

As a first approximation of the variation of all sites (B_1) and of settlements (B_2) within the study area (A) in relation to plant communities (C_1—grassland, C_2—open woodland, C_3—woodland) probabilities are assigned on the basis of observation. Thus,

$$P(A \cdot C_1, B_1) = .18$$
$$P(A \cdot C_2, B_1) = .19$$
$$P(A \cdot C_3, B_1) = .63$$

and

$$P(A \cdot C_1, B_2) = .07$$
$$P(A \cdot C_2, B_2) = .24$$
$$P(A \cdot C_3, B_n) = .69$$

Seeing that the plant community has an effect on the variation of the attribute classes, Plog continues his analysis through the use of one-way and two-way analysis of variance. His findings lead him to conclude that "both the plant community and water resource variables are useful in explaining utilization [distribution of all sites], but only the water resource variable is important in explaining settlement locations" (1972:10). Noting the differential effects of the variables on the two attribute classes, Plog asks,

> Why in one analysis does the community variable seem to be important while in another it does not? These results might best be explained if it were the case that plant communities and water resources significantly covary. In this instance, the plant community variable would mask variation, it seems to account for variation, that was in reality associated with water resources. (1972:10)

The relationship that Plog has informally characterized demonstrates the utility of the screening-off relation. It is in instances such as this, where one variable seems to account sufficiently for the phenomenon, that the screening-off relation supplies a formal means for demonstrating whether a particular partitioning of the reference class really provides more information than an alternative partitioning. Thus, application of the screening-off relation would reveal, in this case, that partitioning the reference class only on the basis of plant community is insufficient to account for variation when a partitioning of the reference class by water resources provides more information about the variation; we would see that water resources screen off plant community from B_2 in A.

The S-R model is descriptive of at least some cases of archaeological research. It would not surprise us if this turns out to be the case when other research programs in anthropology are examined from this perspective. Despite the specific differences between the D-N and S-R models of explanation, both accept as proper the use of scientific induction. Yet the adequacy of that logical approach to understanding can itself be challenged. The next chapter sketches one way in which a cognitive process approach to science might contribute to discussions of the justification of induction.

Seven

Cognition and the Justification of Induction

It is often taken for granted that scientific induction—that is, induction by enumeration—is the "best," most "natural," or most "advanced" form of human thought. In fact, anthropologists have haggled for decades over the question of the primacy of scientific induction in relation to the rationality of "primitive" thought (see Horton and Finnegan 1973). Some maintain that "primitive" thought represents the earliest, least developed stages of intellectual development (Levy-Bruhl 1910; Berry and Dasen 1974). Others argue that all of the varied forms of thought employed by different peoples are rational, while maintaining that scientific induction is in some way more rational than others (Jarvie 1967). Still others argue that all of the varied modes of thought exhibited by humans are essentially equivalent (Radin 1957; Lévi-Strauss 1969).

Anthropological discussions notwithstanding, it is fair to say that the use of scientific induction as a *normative* approach in Western epistemology has gone relatively unquestioned. In fact, the single most telling challenge to the use of scientific induction was raised over two centuries ago by David Hume, first in his *Treatise of Human Nature* (first published in 1739) and later in his *Enquiry Concerning Human Understanding* (first published in 1748). It was Hume's contention that the use of scientific induction is not justifiable because it is based upon the habitual attribution of necessity to customary associations. We learn, that is, to associate A and B based upon past experience. Yet we also come to assume that the association of A and B is a necessary one, one of cause and effect which will continue into the future. Hume found this assumption about the nature of relations in the operational environment anathema. Simply put, Hume asked, How can scientific induction be justified?

Induction and Cognition

Attempts to consider Hume's problem take two general forms (Salmon 1967). First, an attempt is made to justify the use of scientific induction by recourse to induction itself (either scientific induction or another form of induction). This type of approach tends ultimately to result in circularity and has therefore been discredited. Second, some approaches suggest that there is no need to justify the use of induction. After all, scientific induction has worked in the past, is presently working nicely, and will doubtless continue to work in the future. Critics of this type of response rightly point out that it merely begs the question.

Is it not puzzling that a form of inquiry so fundamental to the normative understanding of Western science should prove so difficult to justify? This difficulty can be traced to two considerations. First, the problem is usually presented in a way that leads to the assumption that there is indeed some independent philosophical justification for the use of scientific induction. Second, attempts to justify scientific induction tend to be evaluated, if only implicitly, by the extensive use of scientific induction itself. What must be done is to find some approach to Hume's problem which avoids two pitfalls; a solution must neither beg the question nor fall prey to tautology.

These pitfalls seem, in part, to result from the manner in which Hume's problem is posed. Our intention is to restate the problem, without doing damage to its essence, so that it is less value-laden. We hope our restatement will lead to a process analogous to Kuhn's (1970:3) "changing spectacles," one that will open new avenues of approach to the problem.

The problem restated is: What is scientific induction and should we use it?

The problem now takes the form of a two-part question. The first half of the question inquires after the nature of the *process* that is termed scientific induction, and will entail a modeling approach for an answer. After developing a working model of induction, we inquire into the second half of the question: What particular value should be placed upon the use of scientific induction in epistemology? Briefly, we argue (1) that the term "induction" labels one aspect of a biologically predisposed process of inquiry; (2) that "scientific induction" is an ethnocentric normative application of that aspect of cognition; and (3) that there is no particular virtue in continuing to use a normative scientific induction, as opposed to other forms of induction.

Induction and Inquiry

It is both a biogenetic predisposition and a cognitive imperative for people to impose order upon their reality (see Chapters 1 and 2). At the same time, this reality must be organized so that adaptively salient predictions result from cognitive operations and actions. Earlier we distinguished between reality, which we called the operational environment, and an individual's model of reality, called the cognized environment (see Chapter 2). By prediction we mean that an individual or society must be able to anticipate the results of their actions in order for their individual and collective cognitive systems to operate efficaciously upon the operational environment. The process of operational environment–cognized environment adjustment operates to keep the cognized environment adaptive by imposing redundancy upon information from the operational environment. At the same time it introduces novelty via sensory input. Novelty itself is processed by pattern analysis and, usually, imposition of redundancy. The process of cognition does not stop there. Rather, we perform operations and actions upon the operational environment based on the cognized environment model formed by redundancy patterns. Responses by the operational environment to such operations and actions are monitored for discrepancy. If too great a discrepancy should materialize, the mismatch is coded as novelty, an orientation response is elicited, and the attention is directed at that aspect of the operational environment providing the discrepancy. In all likelihood the discrepancy will be minimized through model reformulation and retesting.

Science is an occidental conceptualization of a genetically predisposed epistemic system (cf. Rensch 1971). One problem with this conceptualization is that it is a simplified model of the system, generally expressed as a tension between "induction" and "deduction." In point of fact, these terms refer to the two phases of the processes of cognition (see Figure 7.1). That is, the information-input–pattern-evaluation phase of inquiry is conceptually isolated and termed "induction," while the model-testing phase is isolated and termed "deduction." The two phases are elevated to the status of "methods" and placed in opposition to each other, often as competing approaches to truth, and access to empirical reality. Once elevated to the status of method, induction becomes "scientific induction" through: (1) normation (i.e., it is constrained and directed by social prescription and proscription to appropriate domains in the operational environment), and (2) phenomenological monophasia (i.e., application directed only at a single—"normal waking"—phase of consciousness. It is thus clear that the common distinction between scientific induction, on the one hand, and scientific deduction, on the other hand, is a highly inaccurate model of scientific cognition. This highly prevalent cognized logic is far more confusing than enlightening in reference to the study of the actual operational logic of science.

A later section argues that science proceeds by way

Figure 7.1
Induction and Deduction as Aspects of Cognitive Adaptation

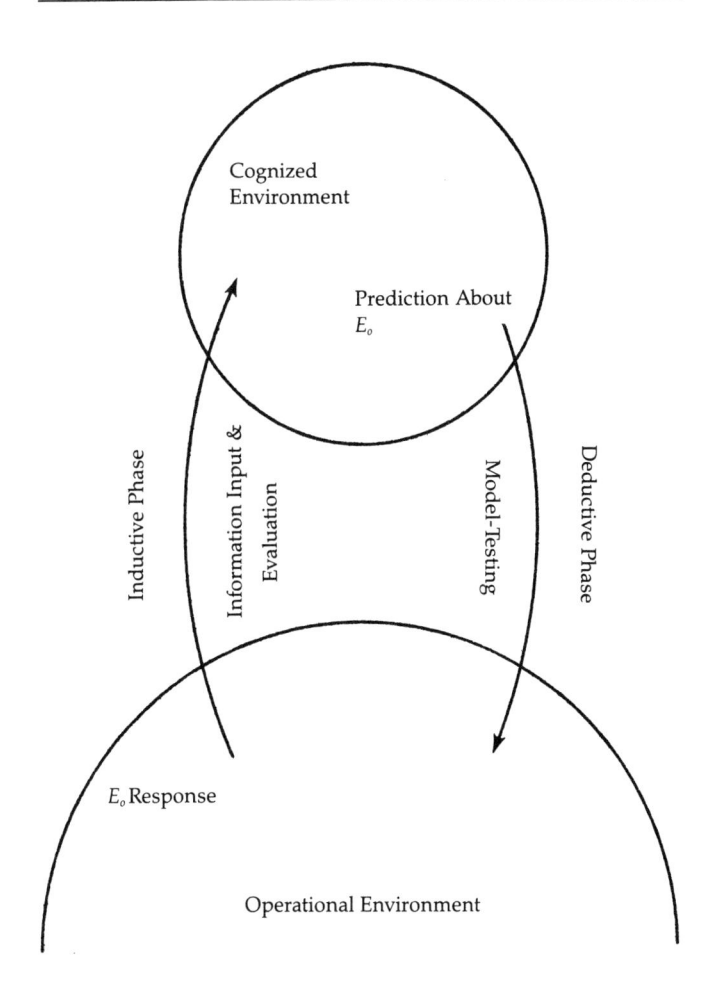

of an alternation between the two poles of operational environment–cognized environment equilibration; the information-input–pattern-evaluation phase (induction), and the model-testing phase (deduction). The factor that results in great empirical efficacy in science is the relative openness of the inductive phase to environment. Not only has induction been made normative, but it includes a proscription against premature closure of theoretical systems. Such systems do, as Kuhn (1970; see also Chapter 4) has shown, ultimately tend to close, but not so quickly or so completely that anomalous information fails to accrue.

Attempts to severely constrain science to a single mode of inquiry, be it either induction or deduction, have led to aberrant and self-defeating approaches to inquiry. "Pure" inductive approaches have led to the "mindless empiricism" exemplified by certain schools of thought in archaeology that contend that all one needs are sufficient artifactual materials and the data "will order themselves." At the other extreme are certain semiotic structuralist approaches that all but ignore contradictory data (see Chapter 6). To put it quite simply, normative truncation of cognition via "pure" induction leads to information overload and confusion, while that via "pure" deduction leads to metaphysics with little or no necessary relation to the operational environment of reference.

Justification of Induction

Thus far we have addressed the question: What is scientific induction? There remains the question: Should we use scientific induction? Our answer is a qualified no, not as the method is commonly and ethnocentrically conceived in textbook science. Earlier chapters have raised several of our reasons for reaching this conclusion.

Scientific Induction Distorts the Role of Deduction in Science

Cognized environment models either comprise or are founded upon other primary models that form the individual's (and the culture's) ontology and epistemology. The construction of theory is of course impossible in the absence of a fundamental set of assumptions about the nature of how we come to know the operational environment. Thus any conceptualization of "pure" induction operating apart from the symbolic function, ontology and epistemology, and model-testing is a distortion of the role of deduction in inquiry and a normative concept with little basis in fact. The importance of the inductive mode is directly proportional to the openness of models and is *not* founded on the absence of models. This position is best described by the Piagetian conception of the tension between assimilation and accommodation. A child assimilates the operational environment when the information input into the cognized environment evokes models requiring no reformulation. On the other hand, when a child manifests accommodation, his or her cognized environment models undergo structural reformulation in relation to operational environment input, in order that he or she may later impose redundancy upon similar information. Extreme assimilation is equivalent to "pure" induction. Either extreme in adults is relatively aberrant (cf. Chapter 4, pp. 70–71).

Scientific Induction Avoids the Issue of Phenomenology

Consciousness is phasically structured. How we experience events initiated within the cognized environment or within the operational environment depends upon the dynamic interaction of subsystems within the nervous system used, from moment to moment, for processing information about those events. Occidental culture tends to cognize different phases of phenomenology as severely distinct rather than as a shifting flow of experience, as do the peoples of traditional Tibet (Trungpa 1976; Bharati 1976; Evans-Wentz 1958; Bloefeld 1974) and Senoi (Stewart 1951, 1953), to note two of many (see Tart 1975). "Science" is typically conceived of as being extremely monophasic, as a system of inquiry applicable only to events perceivable and cognizable during the normal waking phase. Science as generally conceived is what Charles Tart (1975) termed "state dependent." From this perspective, it is clear that scientific induction, forming a part of a monophasic system of inquiry, is itself defined as monophasic in application.

Close adherence to the concept of scientific induction and its implicit monophasia has often led to the contention that "objective" knowledge is possible in science. That is, scientific knowledge is independent of phenomenology and the principles operating to structure the nervous system of the knower (e.g., Popper 1959, 1963). Furthermore, severe monophasia has led most scientists to attend to aspects of the operational environment external to their own organism, as well as to the strictly material and superficial aspects of processes in the cognized environment. Teilhard de Chardin (1959) called this tendency attention to the "without" of things as opposed to the "within" of things. This factor is reflected in the extreme "hallucinophobia" of our social sciences—the fear of one's own phenomenology, or essentially the fear of our own "withinness." An important result of extreme adherence to scientific induction is our almost total inability to come to meaningful terms with alternative sciences (i.e., approaches to empirical knowledge obtained via alternative phases of phenomenology as exemplified by much of the literature pertaining to Tantrism; see Evans-Wentz 1958; Bharati 1976; Trungpa 1976), much of which addresses problems similar to those confronting occidental science.

Induction in the Context of Science

Scientific induction is an occidental normation of fundamental aspects of basic cognitive processes, and scientific induction, as commonly defined as a methodology, should not be viewed as a singularly privileged mode of inquiry because it is thus ethnocentric, epistemologically restrictive, and value-laden. This result is somewhat negative and should be balanced by placing induction into a less restrictive and conceptually more satisfying cognized logic.

Science as Inductive-Deductive Alternation

Science is a process of inquiry that progressively explores the operational environment via a systematic alternation of induction and deduction. *Induction* labels the phase of the cycle of inquiry during which information pertaining to the operational environment is collected and evaluated as being either redundant (i.e., anticipated by, and therefore a verification of an explicit or implicit model) or anomalous (i.e., novel in relation to the model of reference). This phase may result in model reformulation. *Deduction* labels the counterpart of induction, the phase of the cycle of inquiry during which models are initially formulated to give rational coherence to (i.e., explain) phenomena, and in which models are tested for verification. Thus, the two terms label phases of a continuous cycle (continuous unless the cycle is closed due to one or another of the sociocultural factors discussed in Chapters 2, 3, and 4).

Induction and deduction are part of a greater whole, a cyclical process that inevitably involves model formulation and testing. A scientific model may be defined as a set of propositions that describes entities and relations between entities, the existence of at least a portion of which is unverifiable by direct perception. This would seem to present a paradox—at least at the level of cognized logic. Scientific models purport to tell us something about the operational environment and yet are unverifiable by direct reference to the operational environment. The paradox is false, however, and it evaporates when the issues involved are placed in evolutionary perspective. The perceptual unverifiability of many products of human cognition, including scientific model-building, became orthogenetically inevitable the moment an organism became capable of cognizing relations in open space and time more extensive than presented during moment-to-moment "pure" experience. In other terms, the capacity to "time-bind" (Whitehead 1929) or form "plans" (Miller et al. 1960; Count 1974) transcends the boundaries of simple perception and facilitates the construction of an increasingly complex cognized environment that is at once more adaptively versatile and open to operational environment–cognized environment discrepancy. Scientific models may be seen as special instances of models comprising part of a person's or a group's cognized environment. They are special in the sense that they are generally modeled using a sign or formal sign system, are open to (but often do not manifest) total awareness, and may be intersubjectively "objectified" through an expressive symbol system. Yet scientific model-building remains a function of cognition and is subject to the same processual constraints imposed by the structure and function of the nervous system upon cognition generally. Only a partial set of these constraints has usually been cognized and made normative in the enculturation of scientists and regulation of science (see Chapter 4).

Once formulated, scientific models are tested for accuracy of fit with reference to the operational environment. The testing is carried out via hypotheses that predict (or retrodict) some set of relations that will obtain (or has obtained) in the operational environment and that, unlike the model from which they are drawn, can be directly verified with reference to the operational environment. The question remains, How are these hypotheses generated, and how can we be certain they have the power to confirm or disconfirm their parent model? A predictive hypothesis is thought to be the conclusion of a valid deductive argument, the premises of which are all, or some, of the theoretical propositions constituting the parent model.

Propositions 1 through 4 of the model become logical Premises 1 through 4 from which Conclusion A is derived by deduction. Conclusion A then becomes a hypothesis that stands in a special logical relationship to the parent model. Because of its deductive origin, if

Premises 1 through 4 are all true, Conclusion *A* must also be true. Thus, if we look in the operational environment for the anticipated relations and find that these do not in fact obtain in the operational environment, then we can suspect that there is something wrong with one or more of the original premises. But what does it mean if the event or relationship predicted in Conclusion *A* is observed? Does it mean that Premises 1 through 4 must also be true? As Popper (1959, 1963) has shown, the answer is an emphatic "No!". The truth of Conclusion *A* merely confirms that Premises 1 through 4 *may* be true, that there is no reason as yet to believe them false. To put this more formally, the premises may be sufficient but never necessary conditions for Conclusion *A*'s being correct. Conclusion *A* could just as well have been derived from a totally different set of premises (see Chapter 1). This is important because it explains why it is usually necessary to test scientific models repeatedly and under a variety of circumstances for consensual verification (Wimsatt 1981). With each succeeding test, we gain increasing confidence (sometimes misplaced) that the theoretical premises accurately model actual entities and relations in the operational environment.

Generally speaking, the process of model construction is not as simple as depicted above. One or more of the critical first hypotheses will, at best, only partially predict actual events in the operational environment. If these results of experimentation or observation are at variance with predictions deduced from the model, progressive refinement of the model will be required to bring it to fruition. The variant results initially obtained become "inductive" input into the model, feedback drawn into the model through hypothesis testing. The old hypothesis, Conclusion *A*, is first modified so that it becomes a new proposition, Conclusion *A'*, which had it been originally deduced from the model would have predicted the correct observational results. One step further up the inductive ladder, propositions that composed the set of premises from which Conclusion *A* was originally deduced must be changed. We need to determine how these propositions may be changed so that, taken as a new set of premises, 1' through 4', the new Conclusion A' might be deduced. This is the model reformulation phase of induction.

Once the model is modified so that it includes Propositions 1' through 4' and a modified deductive hypothesis, the process of science shifts back to the alternate deductive phase. If this alternation is not made and the result is claimed to be a proper scientific model, an important epistemological responsibility has been shirked and the *post hoc* fallacy is committed (see Chapter 1). It is necessary to generate a new hypothesis using all or some of the modified propositions as premises in a new deductive argument, the conclusion of which, Conclusion B, predicts some event or relation in the operational environment that was not entailed in the prediction tested in the original Conclusion A, or in its modification A'. An alternative strategy, and one that is perfectly valid, is to base Conclusion A' on a body of data independent of the body of data used to test our original Conclusion A. This process will also avoid the *post hoc* fallacy. Either way, the modified model is tested independently, and at the same time new inductive material enters the process (see Figure 7.2).

The process of inquiry continues as before with predictions made by Conclusion B being evaluated for accuracy of fit with the operational environment, empirical input perhaps necessitating modification of Conclusion B to a Conclusion B', and so on. Thus, through a lengthy series of alternations between induction and deduction (as information-processing phase) a body of confirmation is gathered to support a model. Of course, not all scientific models are confirmed. This is merely a model of the process of inquiry commonly used in constructing successful models and rejecting inadequate ones.

This process of inductive-deductive alternation in theory-building is superbly illustrated in Kepler's attempt

Figure 7.2
Inductive/Deductive Alternation in Science*

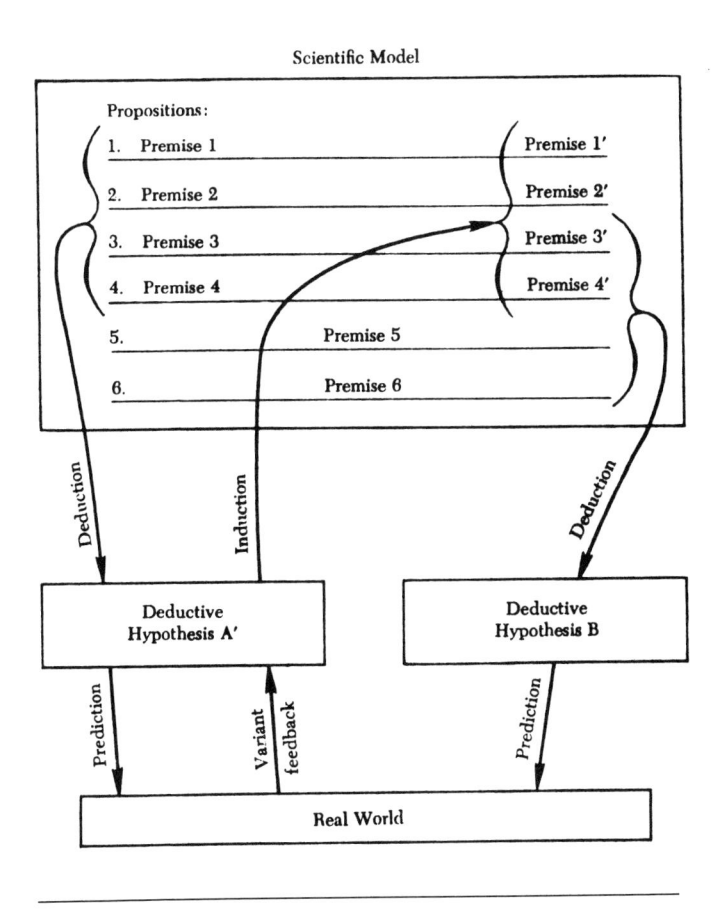

Scientific Model

Propositions:
1. Premise 1
2. Premise 2
3. Premise 3
4. Premise 4
5. Premise 5
6. Premise 6

Premise 1'
Premise 2'
Premise 3'
Premise 4'

Deduction — Induction — Deduction

Deductive Hypothesis A'   Deductive Hypothesis B

Prediction — Variant feedback — Prediction

Real World

*After Laughlin and d'Aquili 1974.

to render the orbit of Mars explicable by simple physical forces, by finding a single geometric figure that describes the planet's orbit. In his account of this work, published in his *New Astronomy* in 1609, Kepler details the steps leading to his ultimate conclusion that Mars has a circular orbit with the sun at its focus. The description he offers is of an eight-year *process* of inductive generalization and deductive testing (see Toulmin and Goodfield 1961).

An equally good example of the process of science in operation from the anthropological literature is a series of studies by John Cove (1973, personal communication) on decision-making among fishing boat captains. Based upon field research among offshore fishermen in Newfoundland, Cove developed a model to account for variance in risk-taking by captains. The model was a formal one, stipulating the various relations among reward structure, technology, and environment which influenced captains' decisions. Had Cove ended his theory-building activities at this point, and claimed that his model accounted for all risk-taking decisions made by fishing captains, he would have incorporated the *post hoc* fallacy into his analysis. However, he went on to test his model twice more, each time revising and generalizing the model for greater empirical coverage. By testing the model, first among inshore salmon fishermen in British Columbia and later among oyster fishermen in Cornwall, England, Cove (1) strengthened the theory-building process by increasing empirical input and (2) discovered that the model in fact could not account for all of the variance in all three social contexts. Cove ended this study by *rejecting* his particular model while retaining confidence in a situational approach to risk-taking decisions. In effect, what Cove's study illustrates is the practical and empirical power of a repetition of the alternation of induction and deduction in scientific theory-building.

Science in Perspective

What is presented above is (1) a cognized logic of, (2) the process of science, which is (3) a special form of human cognitive processing. The process of science is *not* simply a unique by-product of Western philosophical tradition, but is rather a process of inquiry both structured by the biology of neural processing of information and a simplified, although largely conscious, normative application of this process by Western tradition. At its root, science is a special application of general cognitive processes. In order to emphasize the correspondence between the special reconstructed logic of science just presented and a more general model of human cognition, Table 7.1 directly compares the most salient aspects of both models.

The noteworthy advantage of the normative application of a reconstructed logic as presented here is that it can have the effect of socially prescribing constant attention to anomalous information from the operational environment (Pribram's [1971] attention to empirical "paradoxes," for example) and proscribing premature closure of the model-building process. An equally noteworthy disadvantage is the unfortunate loss from general consciousness that the process is a symbolic simplifica-

Table 7.1
Features of General Cognition and Features of Sciencing Compared

General Cognition	Sciencing
1. Initial cognitive model	1. Initial scientific model
2. Generation of expectations about the nature of the operational environment	2. Deduction of hypotheses
3. Behavior on the operational environment based upon expectation and input of perceived operational environment response	3. Hypotheses tested by experimentation or observation, and accumulation of data set
4. Perceptions stored in short-term memory	4. Storage and coding of data set
5. Expectations evaluated in relation to perceived operational environment	5. Confirmation or rejection of hypotheses
6. Physiological modification of cognitive (neural) model	6. Modification of hypotheses and theoretical propositions
7. Generation of expectations about the nature of operational environment	7. Deduction of hypotheses

tion of a complex process and not the complex process itself. Normative applications of simplified alternative applications of the real process of cognition that are perceived by practitioners of a discipline to be outside the "paradigm" are thereafter disattended, or even negatively sanctioned. An excellent and current example from the social sciences is the active disattention on the part of behavioral psychologists to either the field of genetic epistemology or transpersonal psychology.

Several other points may be taken from the present view of science. First, it is interesting that the process of inductive-deductive alternation may be entered at any point in the cycle. The scientist may begin with a model suggested by another theorist and move on to test and further refine the borrowed model. He or she thus enters the cycle at the deductive phase. On the other hand, the scientist may be confronted by a body of data exhibiting a pattern that requires explanation. In constructing a model to account for this pattern, the scientist has entered the cycle at the inductive phase. *What is really epistemologically important is not which phase of the cycle the scientist enters, but rather that the cycle of deductive-inductive alternation in inquiry continues.* Gardner (1978) argues just this point in relation to developmental psychology.

Second, it is always important to remember in dealing with these issues that consciousness, itself, is a function of the brain, and one with a long and complex evolutionary history. Much of cognitive activity normally remains beyond the bounds of consciousness. In modeling the process of science, we have meant to keep our understanding in line with what we know of human cognition. We have *not* meant to imply that all scientists are conscious of every aspect of their own cognition. If practicing scientists were more conscious of the process of science, it would go a long way toward circumventing the epistemological inhibitions imposed by paradigms.

The complexity of our understanding of the operational environment is constrained by the developmental transformations of which the individual cognitive system is capable. Furthermore, the optimal complexity of scientific model-building is constrained by the complexity of the most complex members of a scientific community. As Piaget and his associates have shown, the ontogenetic development of the comprehension of a number of key concepts in science, including consciousness, chance and probability, and causality (Piaget and Inhelder 1969), to mention but a few, follows an invariant progression from simple and concrete to complex and abstract. In addition, it is possible to show that the human capacity for complex cognition is involved in a rapid period of evolution, a process that has very likely not reached a long-term plateau. Karl Pribram (1971, 1978), John Crook (1980), and others have argued that the brain plays a much more constructive role in the interpretation of the world than is usually attributed to it. They, David Bohm (1977, 1978, 1980), and a variety of other researchers (e.g., Wilber 1982) suggest that the recognition and understanding of this provides the opportunity for the development of forms of cognition more complex than now realized. Herbert Koplowitz (cited in Ferguson 1980), for instance, has argued for the potential existence of stages of cognitive development beyond the most advanced Piagetian stages ordinarily found in human cognition. These involve, for example, a stage in which both cause and effect which are not temporally separate can be directly apprehended by an individual. If true, this means that the most complex and viable products of science today as exemplified by quantum theory in physics, mathematical genetics in biology, and genetic epistemology in psychology will one day be interpreted as entirely too simpleminded for serious consideration and will be replaced by models so complex that they will defy comprehension by today's best minds.

Eight

Experience and the Internal Image in Scientific Thinking

The preceding chapters have addressed a number of questions concerning the status and acquisition of knowledge, particularly as they are relevant to anthropology as well as the process of sciencing in general. The approach has been a structural one, applying a developing model in evolutionary structuralism called biogenetic structuralism. A structural approach is advantageous for a number of reasons. It allows the transcendence of both the idiosyncratic boundaries of specific disciplines and the problem of differing levels of analysis that tend to parallel these boundaries. The concept of structure best handles the problem of development on either an evolutionary or an ontogenetic level. Indeed, it may apply to the development of an idea, a culture, or a science as well as an organic process.

Nonetheless, by itself a structural analysis cannot lead to an adequate account of the acquisition of knowledge. Questions remain open, and evidence exists for an expanded model. This evidence exists on a ground somewhat different from that used in earlier discussions.

In almost any scientific endeavor, observation of phenomena and use of the comparative method generally constitute an early stage of development. Later observations are systematized, and controlled observations are made. Some of the material discussed in this chapter involves data only now beginning to enter this mature stage. Yet this material too is vitally important to an understanding of sciencing and of cognition itself.

The questions that concern us here are of the nature and relevance of *experiential knowledge*, the role of the internal image, and the status of figurative knowing.

Kinds of Knowing

There are at least three distinct kinds of knowing: instinctive, logico-mathematical, and experiential (Piaget 1971b). The first is the initial neural models that are hereditarily transmitted and species-specific. These structures are very broadly constrained perceptual-motor schemes, foundations of chreods in Waddington's (1957) sense, guiding later ontogenetic development.

Logico-mathematical knowledge is expressed in structures developing ontogenetically which constitute the organization, first of perceptual-motor adaptive patterns and later the operations of logical thought. These structures arise from the foundations of neurological patterning and they control and direct an individual's adaptive responses to the environment.

To distinguish experiential from the other two forms of knowledge, note that the first is a form of species knowing, partially represented in each phenotype *before* development or experience, and that developmental structures are self-constructed from the *actions* an individual makes in serial encounters with the operational environment. Structures are built up out of patterns in adaptive behavior. Experiential knowledge is information engendered from contact with the environment per se. Experiential or figurative knowledge *encodes properties of the operational environment* which are filtered through and organized within the functioning structural apparatus. These structures appear to exercise logico-mathematical limits upon the acquisition of sensory-based, figurative knowledge, and therefore upon experience itself. Both the recording extant within the organismic system (figurative aspect) and the cognized environment–operational environment interaction are governed by the structures generating the behavior in the encounter. Structures have system-specific sensitivities that play a part in defining the stimulus and the related experience.

Figurative knowing is more heavily influenced by perception than by structural knowing, its properties being spatial, atemporal, and static. Here perception means activity and organization beyond simple sensation, including perceptual activity. Figurative knowledge abstracts properties of objects per se while operative or structural knowledge enriches the object encountered with properties such as order or classification.

Figurative and operative knowing are complementary (Piaget 1971b), comparable to Hebb's (1968) observation that the experience of "seeing" the figurative aspects, the internal image is negatively related to "thinking" *qua* internal verbalizations. Indeed, Hebb sees the vividness of imagery as dependent upon the activation of primary cell assemblies in the sensory cortex. The effect of these cell assemblies is attenuated or often eliminated by activation of secondary or higher cell assemblies involved in "thought" or the association areas of the cortex. This distinction is parallel to the figurative-operative distinction set out above.

Piaget (1971b) posits three kinds of figurative knowing: (1) perception in the presence of an object; (2) imitation with an object either present or absent; and (3) mental imagery in the absence of an object. The third is internally represented or evoked. Hebb points to the higher cell assemblies internally reverberating in the association cortex as the physiological site for evocation, which may or may not spill over into activation of primary cell assemblies. Mental images vary in the degree to which they are purely static. Piagetian research suggests that properties of an internal image that extend beyond a static, atemporal representation are provided by the operative structures.

Figurative knowledge provides a static representation of aspects of the operational environment that is fairly fixed in its perspective. As logical thought develops, the predominance of figurative thought seems to recede. Clinicians long have pointed out the often antagonistic, inhibitory effect that excessively vivid figurative functions have on the operation of logical thought. Between inherited, basic adaptive patterns and the abstract logical thought of the adult human, there is a domain of thought heavily dependent upon internal representations perceived as *images* in various sensory modalities.

Forgotten Levels of Knowledge

There is a stage in the ontogenesis of logical thought heavily dominated by figurative elements. Preoperational thought, beginning with the first indications of internally represented images, is heavily constrained in its functioning by the static properties of figurative thinking. Preceding the onset of logical operations, structures dominated by figurative components are unable to record certain movement and the transformation of events in the operational environment. Because of this dependence on static representations, notions such as conservation of mass, volume, and quantity, aspects of causality, and many other concepts are absent. Intellectual functioning is much more perceptual and experiential since it is the operative structures of later development that enrich reality with much of what an adult perceives. So-called cognitive errors are frequent during this period, particularly in its early stages, and the period is characterized by marked egocentricity in the sense that the child computes events in the operational environment from a position of extreme self-reference. (For example, the sun "follows" him or her as he or she moves; successive rabbits met on a walk are the "same" rabbit.)

It is this preoperational period of development which holds clues to some interesting phenomena in the thinking of adults. The predominance of projection and the dominance of perception and experience of the operational environment by internal events should be noted. The vividness and static character of its images are important. A style of thinking predominates which may occur more frequently than suspected in adult thought and may even be somewhat characteristic of sciencing. In preoperational children this is called *transduction* (Piaget 1971b), referring to the child's tendency to "reason" perceptually from a specific attribute of one object to a specific attribute of another. Since the operational structures generating the distinction between class and member are yet to be constructed, there are no inductive-deductive operations. Comparisons are made on a loose associative basis, rely on similarity of attributes and defy "logical" contradictions. Disconfirming information, though often perceived, is not integrated into the functioning structure. This results in a sort of focused tunnel vision that ignores context. Thought is extremely fluid and symbolic.

As logical operations develop, egocentricity is replaced by an ability to assume alternate perspectives and transduction by more formal logical processes. The importance and experiencing of internal imagery appears to yield to internal verbal communication. However, these functions *are not lost* but incorporated into later adaptive structures. Research on dreams, hypnogogic imagery, and modern therapeutic techniques is now drawing attention to this. For example, dreamlike image material appears to turn on ninety-minute cycles, out of awareness, during our waking hours (Broughton 1976; Klinger 1971). Looking at this information from the perspective of a hierarchical-nested model of cognition, we may ask how much this ongoing imagery influences thinking: What is the structural-functional relationship between the logico-mathematical operations of thought and the preoperational structures upon which they rest?

Two Operational Environments

In this book and elsewhere we have pointed to a Cartesian-like dualism that differentiates between the individual's cognitive construal of reality (cognized environment) and the world outside (operational environment). This is a useful analytic distinction even though it is basically arbitrary. A more satisfactory view is Whitehead's perspective of a multiple-layered sphere encompassing both the cognized environment and the operational environment. It has been helpful to call one level of organization the "cognized environment" and levels both external and internal to that level the "operational environment."

Part of the cognized environment structures are similar to what psychoanalytic writers historically refer to as ego, the seat of logical thought (secondary processes) and of awareness. As such, the internal operational environment constitutes an affective environment, fading from awareness with increasing penetration into its depths and posing an *adaptive problem* for the cognized environment just like that posed by the external operational environment. It is our hypothesis that the ontogenetically most recent levels of structure constitute the major adaptive domain of humans. At the same time, ontogenetically prior structures also constitute an environmental press upon the cognized environment. Cultural variables in the West have reinforced a perspective that emphasizes the external operational environment to the detriment of knowledge and awareness of the internal operational environment. The result is a cognized environment of which we are mostly aware, but not an awareness of its mechanisms or of how it stands between relatively known and a relatively unknown world. Yet, we think, awareness of these is likely to be critical for the fuller understanding of sciencing.

The relatively unknown world of the internal operational environment may continue to exert an effect upon thought, comparable to that exercised by the external operational environment. In fact, we hypothesize that the tendency in normal thought is to confound the two, tending, as in the psychoanalytic notion of projection, to see aspects of *both* in the outside world. Further, as operational structures enrich outside phenomena with their own properties, the earlier, prelogical levels of structure continue to enrich perception and experience of the operational environment through continued functioning outside the bounds of awareness. The greatest influence of this sort may tend to come from the preoperational (prelogical, image-based) levels of organization, biasing our thoughts and perceptions, constraining objectivity in eventually predictable ways.

Where the Data Hide

There are at least three areas of research in cognition to be explored under experimental conditions relevant to this problem. One of these is the notion of *decalages,* or unevenness in development. Substructures develop in clusters at uneven rates (Piaget 1970) only reaching a compatible level and integration at the completion of each major stage in development. For each stage, years pass in states of uneven development (Turiel 1966). A subquestion is: Does a completed stage of development encompass the social as well as the physical environment at the same time? The best guess is that it does not, and that construal of the physical operational environment precedes that of the social and other domains.

A second set of questions asks: What are the effects of variable degrees of integration among these levels and decalages? Is there an optimal range of coordination among these for overall optimal functioning? Do nonoptimal forms of coordination or integration distort or inhibit adaptive functioning? All these factors are likely to affect perceptions of the external operational environment in important ways, and therefore to affect sciencing as well.

Another area needing exploration is the study of cognized environment–external operational environment adaptive fit, particularly as it relates to regressions in functioning (Schroder et al. 1967; Schroder 1971). It has been found among structures measured in relation to the social operational environment that information input affects the functioning of the structures themselves. A critical point for sciencing, particularly important in light of the next hypothesis, is that the U-curve relationship holds for the cognized environment's relationship to the *internal* operational environment as well as for the external operational environment. This is not a novel hypothesis. The Thorndike-Lorge U-curve function of the relationship between anxiety and task performance has long been available in psychology. Our statement is simply in structural terms.

These questions point to an important area. Each involves, in some sense, the relationship of logical thought to its supporting substructures, which, in turn, involve the continuing influence of preoperational levels of organization. It is this level that is most dominated by the figurative image. Decalages may bare this level, nested among an overall structure that appears more advanced. Only in the area of decalage would the more primitive levels constitute the adaptive surface. The degree of coordination among levels, when nonoptimal, may steer the operational structures toward a bias reflected in the organization or content encoded in ontogenetically prior levels. Or there may be a regression from a point internal to the organism toward functioning as if an individual were at a prior level of development. Regressions in the functioning of structures resulting from events in either operational environment tend to be toward prior modes, and the importance of the figurative properties increase. This sort of regression has been called systemic collapse (Rubinstein 1979a).

Structural collapse inclines thought toward more categorical perceptions of the operational environment, increased affective response, and attendant bias on "objective" perception. Among the effects of such a collapse are increased egocentricity (in the Piagetian sense), greater distortion of incoming information toward the internal model, and increased projection of internal content onto the operational environment. All this leads to less "objective" perception, dominated by the bias of earlier structures.

What might this bias be like? It has two potential sources: the structural organizations triggered or exposed and their attendant constraints on functioning, and the content or information encoded within these structures, including an affective bias or valence. The former is characterized by decreasing mobility of internal functioning, a reduction of "depth" in perception of patterns in the operational environment, and an increasingly rigid, dogmatic posture. Important to note in this realm is a somewhat paradoxical increase in transductive reasoning. This is seemingly paradoxical, because transduction evidences another side to the increasing rigidity, a certain fluidity of thinking noted by clinicians in case of pathology—for example, Arieti's pathological thinking in schizophrenic disorders. This fluidity seems to occur within a field characterized by greater overall rigidity. "Loose" associations are permitted, providing a fluid state within a sort of hardening of the categories and lacking the flexibility of functioning which characterizes more advanced structures.

Within each level of structure various forms of information are encoded. Complex structures can hold contradictory and complementary information concerning a single stimulus. The hallmark of a complex structure is the simultaneous and functional coordination of such information in an integrated whole where each bit tends to modify the perspective of the others. The outcome is greater balance in perspective and multidimensional alternative views of that stimulus. Simpler structures, on the other hand, are unable to coordinate contradictory views. The tendency is to isolate disparate information into separate substructures, each internally consistent and characterized by a common value or tone. These lower level structures resemble Jung's complexes (Jung 1969, 1970; Jacobi 1959). In descending into the structures of the internal operational environment, we increasingly encounter more and more fragmented, concrete, uniform, and affectively biased "views" of the eliciting stimulus. Since these structures may often exist in an island-like state, coordinated complex action is minimal. If single clusters are evoked, a uniform, simplistic orientation is activated. If contradictory, mutually inhibitory orientations are evoked, they function either to produce ambivalence, or, in sequence, they result in oscillation.

The form of the "data" encountered is increasingly likely to be a sensory image as preoperational levels are approached. Indeed, even in the deepest reaches, events may be expected to take the form of representation in images as information arises out of the early sensorimotor structures, expressing itself in the codes of the preoperational structures (see Grof 1976, for an example based on therapeutic LSD regressions; many of Jung's works, for nonclinical problems; and such esoteric literature as the *Tibetan Book of the Dead*, for non-Western examples).

What would be the effect on thought of the above hypotheses? What characteristics does thinking display with increasing input or intrusion from the image-based preoperational levels? We expect (1) structural functioning to manifest characteristics of the earlier levels and (2) the content, information, or conflicts encoded in those levels to be "released" or activated, influencing the content of attention and biasing experience. In the former case, thought becomes increasingly one-sided, value-laden, dogmatic, and fixed. Inductive-deductive operations begin to approximate transductive reasoning by association of partial similarity. Affect and thought are increasingly commingled, while perception is increasingly static and unchanging. In the latter case, an increasingly obsessive pull toward specific stimuli or questions can be expected. When two contradictory clusters are exposed at these levels, serious ambivalence or conflict may be engendered in perceptual awareness, or an individual may be aware of one and not the other. In either case, implicit systematic biases are introduced into thinking.

Clues from Clinical Psychology

Suggestive materials about these processes come from clinical psychology and recent attempts to place such observations in a systematic frame. Psychotherapists are increasingly aware of the importance of internal images in manifestations of pathological functioning, they are increasingly likely to use techniques of image manipulation in their treatment strategies, and they are attempting to delineate the relationships between internal images and aspects of verbal thought.

Aaron Beck's (1967) group at the University of Pennsylvania, for example, has been working for years on the clinical problem of depression and developing effective methods for its treatment. In the process of their work, they discovered that depressive thinking operates under a highly determined set of constraints, a "nonlogical" bias in the process itself. At the core of each depressive syndrome, the clinician is likely to find the same cognitive triad: a negative conception of self, a negative interpretation of life experiences, and a nihilistic view of the future. These forms of cognition show certain structural similarities. The information processed by the structure shows a uniform emotional tone, discrepant information is rejected by the structure and does not influence behavior, and the system exhibits the property of "runaway" (Bateson 1979). These characteristics intensify over time and with repeated functioning. Further, the process appears internally "logical," despite its closure to conflicting information. It closely resembles the systemic collapse described above, increasingly exhibiting properties similar to ontogenetically prior stages. Underlying the whole "logical" process lies an image, untested in behavior or thought, which acts as a "premise" for the more "logical" thinking. This image or premise requires only a little practice in order to be seen but generally operates outside the individual's awareness; this is a point of interest. How much of thought, including that of the scientist, is similarly constrained by an unnoticed, unexamined, and constraining image? How much sciencing rests upon unseen and unexamined figurative aspects of thought?

It is within modern psychotherapies that this issue has its clearest form. Gestalt therapy is one of these approaches, as is Grinder and Bandler's (1974) neurolinguistic programming. Probably the most extensive and deepest examination of the role of image is C. G. Jung's comparative method of constructive analysis and the technique of active imagination (Jung 1953). Underlying all these methods is the recognition of the internal and static image *qua* premise for and constraint upon thought. Within these systems the internal image, literally a perceivable picture-in-the-head, acts as the context within which the active operations of thought occur and which determine the parameters of that thought.

Perl's Gestalt therapy and Grinder and Bandler's approach to therapy make use of the empty-chair technique, where feelings are projected outward into space, often creating at least a faint visual image (other senses can be used, the image may exist in all sensory modalities) which can be separately addressed. Therapeutic progress is partially dependent upon developing a "relationship" with this image. Overall it is a process of forging a synthesis of images existing *within* the cognitive system. The technique, then, is a set of operations for externalizing properties of the internal system in a concrete form so that they can be operated upon. The Jungian technique of *active imagination* is similar, developing the ability to make vivid and monitor internal fantasies in order that they be observed. These are then bracketed with mythological material, *à la* the comparative method, to reveal aspects of the underlying structure.

Entering Thought at the Image Level

Erikson's (Erikson et al. 1976) approach to therapy actually allows a cognitive system to internally reequilibrate or readapt. The projected image techniques mentioned above attempt to forge localized synthesis of disparate substructures in order to generate a developmentally more advanced substructural system. Erikson's approach involves operationally tying up the activity of the dominant cerebral hemisphere (e.g., the analytic, logico-mathematical structures) to gain direct access to the nondominant hemisphere (e.g., Gestalt-synthetic operations). This access is via images constructed with words in short phrases, thus helping the structures to organize "new premises" for thought at early levels.

Preoperational levels of thought are dominated by image-like characteristics. This suggests that Erikson's techniques enter structural organization at these levels. The hypothesis of a bilateral shift in hemispheric dominance at the developmental progression from preoperational (age two to seven) to concrete operational thought (age seven to eleven) is similarly suggested here. At any rate, such accessing techniques show the hallmarks of altering markedly the constraints on subsequent logical thought, potentially even freeing the structures themselves for development or more fluid functioning. Two points deserve note: (1) Complex thought appears to depend upon supporting levels of structure that are often revealed in the form of image and symbol; and (2) direct manipulation at the level of and through *image* can result in the modification of thought.

Changes in thought and behavior resulting from image-based techniques are often observed and experienced as "nonlogical" (Watzlawick et al. 1974; Bandler and Grinder 1975). Nonlogical change, in fact, may be *the* predominant mode of behavioral change over the history of our species, as examination of religious communication and ritual (d'Aquili et al. 1979), ethology

(Smith 1979), or anthropological fields such as Shamanism (Eliade 1964; Riddington 1978) suggests. Ritual behavior, in many cases, manipulates both the structures of cognition and the autonomic nervous system, thus increasing the influence and activity of preoperational structures. Under such conditions, communication is usually most effective when metaphorical or symbolic. The result is either the confirmation of the image-based "premises" behind thought or the altering and replacement of old premises with new ones, as in cases of conversion or rites of passage. In either case, the operation takes place at the image level beneath "thought" in a way similar to modern therapeutic techniques.

We hypothesize that much of what is called "experience" and which influences our professional and personal lives enters the cognitive system at the image-processing and image-generation level. This implies that much of what is experienced is located structurally in more developmentally primitive levels, specifically, in or around those levels structured during the preoperational stage of development.

A great deal of the modern cultural "push" for experiential knowledge, as opposed to purely intellectual knowledge, is an argument for the involvement of *all* levels of structural organization in the acquisition of knowledge. The experiential movement can be seen as regressive if it is taken to negate the utility of succeeding structures. Our experience of life, which colors our questions and our intuitive orientation toward potential answers, is heavily influenced by the reaction and activation of lower-level structures. It is a continually operating source of bias at the individual level, affecting the course of scientific inquiry too. At the social level, it has similar effects (as discussed in Chapter 4). The relationship is clearer when paradigms are viewed as standing in the same relation to the scientific community as myth does to archaic cultures and as image-based premises do to individual thought. Paradigms and myth distill instinctive and social wisdom to an overall metaphor guiding and constraining action. At the same time, individual experiential variation is "readout" leaving the common pattern of structure.

The operations of investigation may thus be "artificially" constrained within the boundaries of the static model represented in the image. A seemingly "logical" process of investigation, even on a cultural scale, may be "irrational" when the underlying image is made clear, tested, and relativized vis-à-vis a *potential* universe of discourse. Indeed, much of what constitutes a paradigm may be carried out at the concrete operational level or lower, never reaching the sophistication of formal operations. As Piaget (1970) pointed out, scientific disciplines appear to undergo their own ontogenetic sequence, paralleling those of individual thought. Within this developmental frame, specific scientists may operate abstractly, yet be unaware that their "abstractness" is highly constrained by the images upon which it is based. These paradigmatic images may reflect the developmental level of the science as well as the more personal constraint of experiential-based images of the individual. It is the notion of abstract logical thought, resting upon a bed of highly constraining images, and the need to see, test, and relativize these images, that is the important point. It is the call for a need to identify the context in which logical-scientific thought occurs that is needed.

The Advantages of Images

Images function in ways that balance their negative constraining effects. Images are most likely important in the creative process. It is this process which not only produces novel solutions that advance the course of thought within a paradigm but also correct and even overthrow the paradigm itself. Students of the creative process have long pointed in this direction from outside a structural perspective (e.g., Arieti 1976; Koestler 1964). The place of the intuitive image in sudden insight and creation is clear from the work and experiences of scientists, including for example, Einstein, Watson's double helix, and the discovery of the benezene ring. The integration of dream images into waking life among more primitive peoples is well known in anthropology for its creative as well as traditionally stabilizing effect, similar to the bivalent effects found in ritual (see Stewart for Senoi, and Riddington [1977] for Dunna-Za dream procedures).

Placing the question in a structural context allows us (1) to transcend specific disciplines and (2) develop testable hypotheses based on what is already known about the functional constraints and properties of given levels of cognition. One key element is the creative potential of using some forms of symbolic or transductive logic. It is the somewhat paradoxical nature of simultaneous constraint and the breaking of constraint in symbolic thought that speaks to the creative endeavor.

Beyond placing constraint upon logico-mathematical operations in the form of an unexamined premise, the preoperational image, or the structures in which it occurs, may be the field in which new ideas arise. Creative intuition generally seems to follow a long process of more conventional, verbal wrestling with a problem. The structures of the mind work around, try out, and examine a question from many sides. Yet it is usually during moments of rest or distraction, when sympathetic activity is minimized and parasympathetic activity increases, that the novel solution is "seen," often in the form of a visual image or in a kinesthetic "feeling" for an answer or solution.

Intense intellectual stimulation followed by rest may bring preoperational material into consciousness, one domain setting the context for the other. Within this sequence, the process of lateral association among concepts or premises represented figuratively allows the trying out of novel solutions, outside the constraint of logical-thought *resting on a particular image or premise.* The boundaries or constraints are loosened to permit coordinations of elements or factors not previously "allowed" by the boundaries of the operating "logical" system.

This notion can be formalized a bit in Bateson's concept of *abduction* (Bateson 1979). Bateson defines abduction as the "lateral extension of abstract components of description." He identifies as incidences of abduction: metaphor, dream parable, art, allegory, and the organization of facts in comparative anatomy. While Piaget's notion of transduction focuses on lateral comparisons of elements or events, abduction includes more abstract aspects of pattern in the comparative process. While every incident or event is a pattern of more molecular events, thereby making transductive thought of the five-year-old a homolog of abduction, the existence of succeeding, complementary levels in mature scientists suggests a greater generalization of the process. Scientific abduction may be a homologous development from transduction yet require utilization or contact with that earlier level to function adaptively. That is, not only are the comparisons of patterns in perception necessary, including the initial loose associations, but a refining, internal testing process to establish a criterion of "reasonableness" is necessary. Involved is a shift from comparing specific attributes to a comparison of relations among attributes. Yet the comparison often takes place in perceptive, figurative, nonverbal modes, often antithetical to what common sense describes as "thought."

While the creative process takes place "automatically" for most people, it is apparent that there are more formal, volitional means to activate the process (e.g., in the role played by dreams; Laughlin et al. 1981). The problem reduces in structural terms to developing the technical means of negotiating a transformation in order to entrain different neural substructures or differing states of consciousness in phenomenological terms. The crossing back and forth between levels of structure, or substructures within a level, is only half of the issue. Information also must be transferred from one structure to another, preferably in the awareness of the cognized environment. Techniques of information conservation and integration are then extremely important.

Knowledge and Transformations

Knowledge is constructed from the structures of thought, and its nature is contingent upon the developmental and momentary status of those structures. All forms of knowing involve thought structure or are patterned in various ways. From those levels of structure constituting the cognized environment there are two worlds demanding adaptive response: (1) the world outside and (2) previously developed structures and the information encoded within them. The adaptive problem of the cognized environment is the coordination of knowledge from the two realms of the operational environment.

These adaptations and development itself are intimately intertwined. They are negatively correlated with the tendency to project and therefore to perceive characteristics of one part of the operational environment within the other. The preoperational levels both constrain the structural adaptations of thought and constitute the potential locus of creative synthesis. Both structures themselves and the information encoded within earlier levels operate outside the range of awareness of most people, thereby maximizing the tendency toward projection. This process operates in human thought in general and is therefore intimately involved in scientific thought despite what are taken as the formal safeguards and formal theories of science.

In a scientific tradition whose hallmark is the development and refinement of theory, of observation, and of the external technology by which it is carried out, a self-conscious awareness of the nature of the development and transformations requisite for the continued acquisition of knowledge is of key importance. This amounts in part to the development of control over the bias and internal technology of the scientist herself or himself. It requires the internalization of the concept of technology and applies it to the scientist, thus blurring the artificial boundary between observer and observed. It has the potential of making us specify the sources of observational error and bias. As important, development and exploration of these processes argues for increasing the scope and range of scientific observation by requiring the inclusion of phenomenological data along with structural and behavioral data in scientific reports. This further suggests the particular importance of knowledge about the properties of figurative knowing. Since the data encountered within this form of knowing are in sensory images, increased study of the nature and role of image and symbolic interpretation in science is essential. At present, the strongest methods for such study come from clinical fields, the literature and techniques of Shamanism in other cultures, and from the esoteric sciences of the East. Indeed, it is in Jungian psychotherapy within Western culture and Mahayana Buddhist techniques that the most sophisticated such knowledge and technology can now be found. It is also incumbent upon scientists to understand and develop such techniques within our own cultural tradition and synthesize them in ways compatible with our research methods. And there must equally be a development of methods and theories that are compatible with such bodies of knowledge. The greatest frontier in the advancement of science rests, we think, upon understanding, controlling, and developing its most critical and complex instrument, the scientist herself or himself.

Bibliography

Ager, T., J. Aronson, and R. Weingard. 1974. "Are bridge laws really necessary?" *Noûs* 8:119–34.

Almeder, R. 1973. "Science and idealism." *Philosophy of Science* 40(2):242–54.

———. 1975. "Fallibilism and the ultimate irreversible opinion." *American Philosophical Quarterly, Monograph Series: Studies in Epistemology,* Monograph No. 9:33–54.

Arieti, S. 1974. *Interpretation of Schizophrenia.* New York: Basic Books.

———. 1976. *The Intrapsychic Self: Feeling and Cognition in Health and Mental Illness.* New York: Basic Books.

Aronson, J. n.d. "Understanding without foresight." Manuscript.

———. 1969. "Explanation without laws." *Journal of Philosophy* 17:541–57.

———. 1984. *A Realist Philosophy of Science.* New York: St. Martins Press.

Bandler, R., and J. Grinder. 1975. *Structure of Magic, Volume 1.* Palo Alto, Calif.: Science and Behavior Books.

Barnett, H. G. 1953. *Innovation.* New York: McGraw-Hill.

Bateson, G. 1935. "Culture contact and schismogenesis." *Man* 35 (reprinted in Bateson 1972:61–72).

———. 1941. "Experiments in thinking about observed ethnological material." *Philosophy of Science* 8 (reprinted in Bateson 1972:73–87).

———. 1972. *Steps to an Ecology of Mind.* New York: Chandler Publishing Co.

———. 1979. *Mind and Nature a Necessary Unity.* New York: E. P. Dutton.

Beck, A. 1967. *Depression: Causes and Treatment.* Philadelphia: University of Pennsylvania Press.

———. 1973. *Diagnosis and Management of Depression.* Philadelphia: University of Pennsylvania Press.

Belshaw, C. 1965. *Traditional Exchange and Modern Markets.* Englewood Cliffs, N.J.: Prentice-Hall.

Berry, J., and P. Dasen, eds. 1974. *Culture and Cognition: Readings in Cross-cultural Psychology.* London: Methuen.

Beth, E., and J. Piaget. 1966. *Mathematical Epistemology and Psychology.* Dordrecht, Holland: D. Reidel.

Bharati, A. 1976. *The Tantric Tradition*. Santa Barbara, Calif.: Ross-Erickson.
Binford, L. 1965. "Archaeological systematics and the study of culture process." *American Antiquity* 31:203–10.
Binford, S., and L. Binford. 1966. "A preliminary analysis of functional variability in the Lousterian of Lavallors facies." *American Antiquity* 31:238–95.
———, eds. 1968. *New Perspectives in Archaeology*. Chicago: Aldine.
Blackburn, T. 1971. "Sensuous-intellectual complementarity in science." *Science* 172:1003–7.
Blalock, H. M. 1969. *Theory Construction*. Englewood Cliffs, N.J.: Prentice-Hall.
Blofeld, J. 1974. *Beyond the Gods: Taoist and Buddhist Mysticism*. London: Allen & Unwin.
Bohm, D. 1977. "Science as perception-communication." In *The Structure of Scientific Theories*, 2d ed., ed. F. Suppe. Urbana: University of Illinois Press.
———. 1978. "The enfolding-unfolding universe: A conversation with David Bohm." In *The Holographic Paradigm and Other Paradoxes* (1982), ed. K. Wilber. Boulder, Colo.: Shambhala.
———. 1980. *Wholeness and the Implicate Order*. London: Routledge & Kegan Paul.
Bordes, F. 1972. *A Tale of Two Caves*. New York: Harper & Row.
Boyd, R. 1972. "Determinism, laws, and predictability in principle." *Philosophy of Science* 39:431–50.
Braithwaite, R. B. 1953. *Scientific Explanation*. New York: Harper Torchbooks.
Broughton, R. 1976. "Biorythmic variations in consciousness and psychological functions." *Canadian Psychological Review* 16(4):217–39.
Buckley, W. 1967. *Sociology and Modern Systems Theory*. Englewood Cliffs, N.J.: Prentice-Hall.

Chagnon, N. 1977. *Yanomamo: The Fierce People*. 2d ed. New York: Holt, Rinehart & Winston.
Chagnon, N., and W. Irons, eds. 1979. *Evolutionary Biology and Human Social Behavior*. Belmont, Calif.: Duxbury Press.
Chaney, R. 1972. "Scientific inquiry and models of sociocultural data patterning: An epilogue." In *Models in Archaeology*, ed. D. L. Clarke. London: Methuen.

Chaney, R., K. Morton, and T. Moore. 1972. "On the entangled problems of selection and conceptual organization." *American Anthropologist* 72:221–30.

Chapple, E. 1970. *Culture and Biological Man.* New York: Holt, Rinehart & Winston.

Chapple, E. and C. Coon. 1942. *Principles of Anthropology.* New York: H. Holt & Co.

Chenall, R. 1971. "Positivism and the collection of data." *American Antiquity* 36:372–73.

Chomsky, N. 1965. *Aspects of the Theory of Syntax.* Cambridge, Mass.: M.I.T. Press.

———. 1972. *Language and Mind.* 2d ed. New York: Harcourt Brace Jovanovich.

Coffman, T. L. 1971. "Personality structure, involvement, and the consequences of taking a stand." In *Personality Theory and Information Processing,* ed. H. Schroder and P. Suefeld. New York: Ronald Press.

Cole, M., and S. Scribner. 1975. "Theorizing about the socialization of cognition." *Ethos* 3:249–68.

Conklin, H. 1964. "Ethnogeneological method." In *Explorations in Cultural Anthropology,* ed. W. Goodenough. New York: McGraw-Hill (reprinted in Tyler 1969).

Count, E. W. 1973. *Being and Becoming Human.* New York: Van Nostrand.

———. 1974. " 'Homination': Organism and process." In *Sonderdruck aus Bevolkerungsbiologie.* Stuttgart: Gustav Fisher.

———. n.d. "The idea of the biogram (prospectus)." Manuscript.

Cove, J. 1973. "Hunters, trappers, and gatherers of the sea: A comparative study of fishing strategies." *Journal Fisheries Research Board of Canada* 30:249–59.

Crook, J. 1980. *The Evolution of Human Consciousness.* Oxford: Oxford University Press.

d'Aquili, E. 1972. "The biopsychological determinants of culture." Addison-Wesley Modular Publications, Module 13, 1972:1–29.

d'Aquili, E., and G. Mihalik. 1977. "Malnutrition: Its effect on psychological development and cultural evolution." In *Malnutrition, Behavior and Social Organization,* ed. L. Greene. New York: Academic Press.

d'Aquili, E., C. D. Laughlin, and J. McManus. 1979. *The Spec-*

trum of Ritual: A Biogenetic Structural Analysis. New York: Columbia University Press.

Dasen, P. 1972. "Cross-cultural Piagetian research: A summary." *Journal of Cross-cultural Psychology* 3:23–40.

──────. 1977. "Introduction." In *Piagetian Psychology,* ed. P. Dasen. New York: Gardner Press.

Deetz, J. 1965. *The Dynamics of Stylistic Change in Arikara Ceramics.* Series in Anthropology No. 4. Urbana: University of Illinois.

Denham, W. 1971. "Energy relations and some basic properties of primate social organization." *American Anthropologist* 73:77–85.

Durkheim, E. 1950. *Rules of Sociological Method.* 8th ed. New York: Free Press.

Einstein, A. 1949. *Albert Einstein: Philosopher-Scientist.* Evanston, Ill.: Library of Living Philosophers.

Eliade, M. 1964. *Myth and Reality.* New York: Harper & Row.

Erikson, M., E. Rossi, and S. Rossi. 1976. *Hypnotic Realities: The Induction of Clinical Hypnosis and Forms of Indirect Suggestion.* New York: Irvington.

Evans-Wentz, W. Y. 1958. *Tibetan Yoga and Secret Doctrines.* New York: Oxford University Press.

Feldman, C., B. Lee, J. McLean, D. Pillemer, and J. Murry. 1974. *The Development of Adaptive Intelligence.* San Francisco: Jossey-Bass.

Ferguson, M. 1980. *The Aquarian Conspiracy: Personal and Social Transformation in the 1980s.* Los Angeles: Tarcher.

Festinger, L. 1956. *A Theory of Cognitive Dissonance.* Stanford, Calif.: Stanford University Press.

Feyerabend, P. 1965. "Reply to criticism." In *Boston Studies in the Philosophy of Science,* ed. R. Cohen and M. Wartofsky. New York: Humanities Press.

Fishbein, H. D. 1976. *Evolution, Development, and Children's Learning.* Pacific Palisades, Calif.: Goodyear.

Flavell, J. 1963. *The Developmental Psychology of Jean Piaget.* Princeton: Van Nostrand.

Frake, C. 1962. "The ethnographic study of cognitive systems." In *Anthropology and Human Behavior,* ed. T. Gladwin and W. Sturtevant. Washington, D.C.: Anthropological Society of Washington.

Fritz, J., and F. Plog. 1970. "The nature of archaeological explanation." *American Antiquity* 35:405–12.

Furth, H. 1969. *Piaget and Knowledge: Theoretical Foundations.* Englewood Cliffs, N.J.: Prentice-Hall.

Gardner, H. 1978. *Developmental Psychology: An Introduction.* Boston: Little, Brown, & Co.

———. 1981. *The Quest for Mind.* 2d ed. New York: Random House.

Giere, R. 1976. "Empirical probability, objective statistical methods, and scientific inquiry." In *Foundations of Probability and Statistics and Statistical Theories in Science,* ed. C. Hooker and W. Harper. Dordrecht, Holland: D. Reidel.

Gilligan, C. 1982. *In a Different Voice: Psychological Theory and Women's Development.* Cambridge, Mass.: Harvard University Press.

Ginsberg, H., and S. Opper. 1969. *Piaget's Theory of Intellectual Development: An Introduction.* Englewood Cliffs, N.J.: Prentice-Hall.

Greeno, J. 1970. "Evaluation of statistical hypotheses using information transmitted." *Philosophy of Science* 37:279–93.

Grinder, J., and R. Bandler. 1974. *Trance Formations: Neurolinguistic Programming and the Structure of Hypnosis.* Moab, Utah: Real People.

Grof, S. 1976. *Realms of Human Unconscious: Observations from LSD Research.* New York: Dutton.

Haan, N., M. Smith, and J. Block. 1968. "The moral reasoning of young adults: Political-social behavior, family background and personality correlates." *Journal of Personality and Social Psychology* 10:185–201.

Hammel, E., ed. 1965. *Formal Semantic Analysis. American Anthropologist,* Special Publication 67.

Hanson, N. R. 1958. *Patterns of Discovery.* Cambridge: Cambridge University Press.

Harvey, O., D. Hunt, and S. Schroder. 1961. *Conceptual Systems and Personality Organization.* New York: John Wiley & Sons.

Hastorf, S., and E. Cantril. 1954. "They saw a game: A case study." *Journal of Abnormal and Social Pyschology* 49:129–34.

Hebb, D. O. 1968. "Concerning imagery." *Psychological Review* 75:466–77.

Hempel, C. 1962. "Deductive-nomological vs. statistical explanation." In *Minnesota Studies in the Philosophy of Science*, volume III, ed. H. Feigl and M. Scriven. Minneapolis: University of Minnesota Press.

———. 1965. *Aspects of Scientific Explanation*. New York: Free Press.

Hill, C. 1974. "Graduate education in anthropology: conflicting role identity in fieldwork." *Human Organization* 33:408–12.

Hinde, R. 1970. *Animal Behavior*. New York: McGraw-Hill.

Horton, R. 1967. "African traditional thought and Western science." *Africa* 37:50–71.

Horton, R., and R. Finnegan, eds. 1973. *Modes of Thought*. London: Faber & Faber.

Hufford, D. 1982. *The Terror That Comes in the Night: An Experience-centered Study of Supernatural Assault Traditions*. Philadelphia: University of Pennsylvania Press.

———. 1984. "The supernatural and the sociology of knowledge: Explaining academic belief." *New York Folklore Quarterly*, in press.

Hume, D. 1739. *A Treatise on Human Nature*. New York: Collins (1976).

———. 1748. *An Enquiry Concerning Human Understanding*. LaSalle, Ill.: Open Court (1966).

Hunter, I. M. L. 1977. "Mental calculation." In *Thinking: Readings in Cognitive Science*, ed. P. Johnson-Laird and P. Wason. Cambridge: Cambridge University Press.

Hymes, D. H. 1974. *Foundations in Sociolinguistics: An Ethnographic Approach*. Philadelphia: University of Pennsylvania Press.

Inhelder, B., H. Sinclair, and M. Bovet. 1974. *Learning and the Development of Cognition*. Cambridge, Mass.: Harvard University Press.

Jacobi, J. 1959. *Complex-Archetype-Symbol in the Psychology of C. G. Jung*. Princeton: Princeton University Press.

Jarvie, I. C. 1967. *The Revolution in Anthropology*. London: Routledge & Kegan Paul.

Jeffery, R. 1969. "Statistical explanation and statistical inference." In *Essays in Honor of Carl G. Hempel*, ed. N. Rescher. Dordrecht, Holland: D. Reidel.

Jung, C. 1953. *Development of Personality*. Princeton: Princeton University Press.

———. 1969. *Mandala Symbolism.* Princeton: Princeton University Press.

———. 1970. *Analytical Psychology: Its Theory and Practice.* New York: Random House.

Kagan, J., R. Kearsley, P. Zelazo, and C. Minton. 1978. *Infancy: Its Place in Human Development.* Cambridge, Mass.: Harvard University Press.

Kaplan, A. 1964. *The Conduct of Inquiry.* Scranton, Pa.: Chandler.

Kleinbaum, D., and L. Kupper. 1978. *Applied Regression Analysis and Other Multivariable Methods.* North Scituate, Mass.: Duxbury Press.

Klinger, E., ed. 1981. *Imagery,* volume 2, *Concepts, Results, and Applications.* Minneapolis: University of Minnesota Press.

Koertge, N. 1975. "Popper's metaphysical research program for the human sciences." *Inquiry* 18:437–62.

Koestler, A. 1964. *Thieves in the Night.* New York: Macmillan Co.

———. 1971. *The Case of the Midwife Toad.* New York: Random House.

Kohlberg, L. 1963. "Moral development and identification." In *Child Psychology,* ed. H. Stevenson. Chicago: University of Chicago Press.

———. 1969. "Stage and sequence: The cognitive developmental approach to socialization." In *Handbook of Socialization Theory and Research,* ed. D. Goslin. Chicago: Rand McNally & Co.

Kohlberg, L., and C. Gilligan. 1972. "The adolescent as philosopher: The discovery of self in a post-conventional world." *Daedalus* 100:1051–86.

Kuhn, T. S. 1962. *The Structure of Scientific Revolutions.* Chicago: University of Chicago Press.

———. 1963. "The function of dogma in scientific research." In *Scientific Change,* ed. A. Crombie. London: Heinemann & Co.

———. 1970. "Logic of discovery or psychology of research." In *Criticism and the Growth of Knowledge,* ed. I. Lakatos and A. Musgrave. Cambridge: Cambridge University Press.

———. 1974. "Second thoughts on paradigms." In *The Structure of Scientific Theories,* ed. F. Suppe. Urbana: University of Illinois Press.

Lancy, D., and A. Strathern. 1981. " 'Making Twos': Pairing as an alternative to the taxonomic mode of representation." *American Anthropologist* 83:773–95.

Langer, J. 1969. *Theories of Development.* New York: Holt, Rinehart & Winston.

Laughlin, C. D. 1974. "Deprivation and reciprocity." *Man* 9:380–96.

———. 1978. "Adaptation and exchange in So: A diachronic study of deprivation." In *Extinction and Survival in Human Populations,* ed. C. D. Laughlin and I. Brady. New York: Columbia University Press.

Laughlin, C. D., and I. Brady, eds. 1978. *Extinction and Survival in Human Populations.* New York: Columbia University Press.

Laughlin, C. D., and E. d'Aquili. 1974. *Biogenetic Structuralism.* New York: Columbia University Press.

Laughlin, C. D., J. McManus, and C. Stephens. 1981. "A model of brain and symbol." *Semiotica* 33:211–36.

Laughlin, C. D., and C. Stephens. 1980. "Symbolism, canalization, and P-structure." In *Symbol as Sense,* ed. M. Foster and S. Brandes. New York: Academic Press.

LeBlanc, S. 1975. "Micro-seriation: A method for fine chronological differentiation." *American Antiquity* 40:22–38.

Lévi-Strauss, C. 1966. *The Savage Mind.* Chicago: University of Chicago Press.

———. 1969. *Elementary Structures of Kinship.* Boston: Beacon Press.

Levin, M. 1973. "On explanation in archaeology: A rebuttal to Fritz and Plog." *American Antiquity* 38:387–95.

Levy-Bruhl, L. 1910. *Primitive Mentality.* New York: Harper & Row.

Lex, B. 1979. "The neurobiology of ritual trance." In *The Spectrum of Ritual,* by E. G. d'Aquili et al. New York: Columbia University Press.

Lomnitz, L., and J. Fortes. 1981. "Socialization of scientists: The internalization of a myth." Paper presented to the 6th annual meeting of the Society for Social Studies of Science, November 2–5.

Lomnitz, L., and C. Lomnitz. 1977. "La creacíon científica." *Pensamiento Universitario,* no. 3.

Lounsbury, F. 1964. "A formal account of the Crow- and Omaha-type kinship terminologies." In *Explorations in Cultural Anthropology,* ed. W. H. Goodenough. New York: McGraw-Hill.

Luria, A. R. 1976. *The Working Brain*. New York: Basic Books.

Mahoney, M. J. 1976. *Scientist as Subject: The Psychological Imperative*. Cambridge, Mass.: Ballinger.
_____. 1979. "Psychology of the scientist." *Social Studies of Science* 9:349–75.
Mahoney, M. J., A. Kazdin, and M. Kenigsberg. 1978. "Getting published." *Cognitive Therapy and Research* 2:69–70.
Martindale, C. 1977. "Comment." *Current Anthropology* 18:473.
Masterman, M. 1970. "The nature of a paradigm." In *Criticism and the Growth of Knowledge*, ed. I. Lakatos and A. Musgrave. Cambridge: Cambridge University Press.
McManus, J. 1975. "Psychopathology as errors in cognitive adaptation." Paper presented at the annual meeting of the American Anthropological Association, San Francisco.
_____. 1979a. "Ritual and ontogenetic development." In *The Spectrum of Ritual*, by E. G. d'Aquili et al. New York: Columbia University Press.
_____. 1979b. "Ritual and human social cognition." In *The Spectrum of Ritual*, by E. G. d'Aquili et al. New York: Columbia University Press.
Miller, G. A., E. Galanter, and K. Pribram. 1960. *Plans and the Structure of Behavior*. New York: Holt, Rinehart & Winston.
Missner, M. 1970. *Chomsky's Concept of Implicit Knowledge*. Doctoral Dissertation, University of Chicago.
Mueller, J., K. Schuessler, and H. Costner. 1977. *Statistical Reasoning in Sociology*. 2d ed. Boston: Houghton Mifflin Co.

Nagel, E. 1961. *The Structure of Science*. New York: Harcourt, Brace & World.
Neisser, U. 1967. *Cognitive Psychology*. New York: Appleton-Century-Crofts.
_____. 1976. *Cognition and Reality: Principles and Implications of Cognitive Psychology*. San Francisco: Freeman.
Newell, A. 1973. "You can't play 20 questions with nature and win." In *Visual Information Processing*, ed. W. G. Chase. New York: Academic Press.

Odum, E. 1971. *Fundamentals of Ecology*. 3d ed. Philadelphia: W. B. Saunders Co.
Ornstein, R. 1978. *The Mind Field*. New York: Grossman.

Paredes, J., and M. Hepburn. 1976. "The split brain and

the culture-and-cognition paradox." *Current Anthropology* 17:121–27.
Petrie, W. M. F. 1899. "Sequences in prehistoric remains." *Journal of the Royal Anthropological Institute* 29:295–301.
Phillips, J. 1969. *The Origins of Intellect: Piaget's Theory*. San Francisco: Freeman.
Piaget, J. 1969. *Genetic Epistemology*. New York: Columbia University Press.
———. 1970. *Structuralism*. New York: Basic Books.
———. 1971a. *Biology and Knowledge*. Chicago: University of Chicago Press.
———. 1971b. *Mental Imagery in the Child*. London: Routledge & Kegan Paul.
———. 1972. "Intellectual evolution from adolescence to adulthood." *Human Development* 15:1–12.
———. 1973. *Main Trends in Psychology*. London: Allen & Unwin.
———. 1974. *Understanding Causality*. New York: W. W. Norton.
———. 1977. *The Development of Thought: Equilibration of Cognitive Structures*. New York: Viking Press.
Piaget, J., and B. Inhelder. 1969. *The Psychology of the Child*. New York: Basic Books.
Pinxten, R. 1981. "Observation in anthropology: Positivism and subjectivism combined." *Communication and Cognition* 14:57–83.
Plog, F. 1972. "Settlement distribution in the Chevelon drainage: A preliminary report." Paper read at the 2d annual meeting of the Southwestern Anthropological Research Group, Tucson, Arizona, March.
———. 1973. "Laws, systems of law, and the explanation of observed variation." In *The Explanation of Culture Change*, ed. C. Renfrew. London: Duckworth & Co.
———. 1974. *The Study of Prehistoric Change*. New York: Academic Press.
———. 1976. "Ceramic analysis." In *Chevelon Archaeological Survey*, Monograph No. 2, ed. F. Plog, J. Hill, and D. Read. UCLA Archaeological Survey.
Plog, F., and J. Hill. 1971. "Explaining variability in the distribution of sites." In *The Distribution of Prehistoric Population Aggregates*, ed. G. Gumerman. Proceedings of the First Southwestern Anthropological Research Group. Prescott College Anthropological Papers, No. 1.

---. 1972. "The southwestern anthropological research group: Revisions of the research design—1972." In *Proceedings of the Second Annual Meeting of the Southwestern Anthropological Research Group,* ed. G. Gumerman. Prescott College Anthropological Reports, No. 3.

Popper, K. 1959. *The Logic of Scientific Discovery.* New York: Harper & Row.

---. 1963. *Conjections and Refutations.* New York: Harper & Row.

Pribram, K. 1971. *Languages of the Brain.* Englewood Cliffs, N.J.: Prentice-Hall.

---. 1978. "What the fuss is all about." Reprinted in *The Holographic Paradigm and Other Paradoxes* (1982), ed. K. Wilber. Boulder, Colo.: Shambhala.

Price-Williams, D. R., ed. 1969. *Cross-cultural Studies.* Baltimore: Penguin Books.

Quine, W. 1960. *Word and Object.* New York: MIT/Wiley.

---. 1964. "On simple theories of a complex world." In *Form and Strategy in Science,* ed. J. Gregg and F. Harris. Dordrecht, Holland: D. Reidel.

---. 1969. *Ontological Relativity and Other Essays.* New York: Columbia University Press.

Radford, J., and A. Burton. 1974. *Thinking: Its Nature and Development.* London: John Wiley & Sons.

Radin, P. 1957. *Primitive Man as Philosopher.* New York: Dover.

Read, D. 1974. "Some comments on typologies in archaeology and an outline of methodology." *American Antiquity* 39:216–42.

Reichenbach, H. 1938. *Experience and Prediction.* Chicago: University of Chicago Press.

Rensch, B. 1959. *Evolution Above the Species Level.* New York: Columbia University Press.

---. 1971. *Biophilosophy.* New York: Columbia University Press.

Riddington, R. 1977. "Monsters and the anthropologist's reality." Paper presented at "Sasquatch and Other Phenomena," Vancouver, Canada.

---. 1978. "Sequence and hierarchy in cultural experience." Paper presented at the annual meeting of the Canadian Ethnological Society, London, Ont., Canada.

Rosenthal, R. 1963. "On the social psychology of the psychological experiment: The experimenter's hypothesis as an unintended determinant of experimental results." *American Scientist* 51:268–83.

Rossi, I. 1974. *The Unconscious in Culture.* New York: Dutton.

Rubinstein, R. A. 1975. "Reciprocity and resource deprivation among the urban poor of Mexico City." *Urban Anthropology* 4:251–64.

————. 1976. *Cognitive Development and the Acquisition of Semantic Knowledge in Northern Belize.* Doctoral Dissertation, State University of New York at Binghamton.

————. 1979a. "The cognitive consequences of bilingual education in Northern Belize." *American Ethnologist* 6:583–601.

————. 1979b. "On 'Culture and sociobiology.'" *American Anthropologist* 81:653–54.

————. 1984. "Epidemiology and anthropology: Notes on science and scientism." *Communication and Cognition.* In press.

Rubinstein, R. A., and B. Donaldson. 1975. "Reappraising anthropological explanation: The archaeological case." Paper presented at the 15th annual meeting of the Northeast Anthropological Association, Potsdam, New York.

Rubinstein, R. A., and C. D. Laughlin. 1977. "Bridging levels of systemic organization." *Current Anthropology* 18:459–81.

Rubinstein, R. A., and S. Tax. 1981. "Literacy, evolution, and development." In *The Spirit of the Earth: A Teilhard Centennial Celebration,* ed. J. Perlinski. New York: Seabury Press.

Rudner, R. 1966. *Philosophy of Social Science.* Englewood Cliffs, N.J.: Prentice-Hall.

Russell, B. 1956. "The philosophy of logical atomism." In *Logic and Knowledge,* ed. R. Marsh. New York: Macmillan Co.

Sade, D. 1972. "Sociometrics of *Macaca Mulatta* I: Linkages and cliques in grooming matrices." *Folia Primatologica* 18:196–223.

Salmon, W. C. 1967. *The Foundations of Scientific Inference.* Pittsburgh: University of Pittsburgh Press.

———. 1971. *Statistical Explanation and Statistical Relevance.* Pittsburgh: University of Pittsburgh Press.
Saltz, E. 1971. *The Cognitive Basis of Human Learning.* Homewood, Ill.: Dorsey Press.
Sanday, P. 1968. "The psychological reality of American English kinship terms: An information processing account." *American Anthropologist* 70:508–23.
Schneider, H. K. 1974. *Economic Man.* New York: Free Press.
Schneour, E. 1974. *The Malnourished Mind.* Garden City, N.Y.: Doubleday & Co.
Schroder, H. 1971. "Conceptual complexity and personality organization." In *Personality Theory and Information Processing,* ed. H. Schroder and P. Suefeld. New York: Ronald Press.
Schroder, H., M. Driver, and S. Streufert. 1967. *Human Information Processing.* New York: Holt, Rinehart & Winston.
Schroder, H., and P. Suefeld, eds. 1971. *Personality Theory and Information Processing.* New York: Ronald Press.
Seligman, M., and J. Hager. 1972. *Biological Boundaries of Learning.* New York: Appleton-Century-Crofts.
Shapere, D. 1964. "The structure of scientific revolutions." *Philosophical Review* 73:383–94.
Simpson, E. 1974. "Moral development research." *Human Development* 17:81–106.
Skinner, B. F. 1953. *Science and Human Behavior.* New York: Macmillan Co.
———. 1969. *Contingencies of Reinforcement: A Theoretical Analysis.* New York: Appleton-Century-Crofts.
Smith, W. J. 1979. "Ritual and the ethology of communicating." In *The Spectrum of Ritual,* by E. G. d'Aquili et al. New York: Columbia University Press.
Spaulding, A. 1960. "The dimensions of archaeology." In *Essays in the Science of Culture: In Honor of Leslie A. White,* ed. G. Dole and R. Carneiro. New York: T. Y. Crowell.
———. 1968. "Explanation in archaeology." In *New Perspectives in Archaeology,* ed. S. Binford and L. Binford. Chicago: Aldine.
Spradley, J. 1970. *You Owe Yourself a Drunk.* Boston: Little, Brown & Co.
Spradley, J., and D. McCurdy, eds. 1972. *The Cultural Experience: Ethnography in Complex Society.* Chicago: SRA.

Stewart, K. 1951. "Dream theory in Malaya." *Complex* 6:21–34.
────── . 1953. "Culture and personality in two primitive groups." *Complex* 8:2–23.
Straight, H. S., ed. 1974. *The Psychological Bases of Linguistic and Other Rules*. Binghamton: State University of New York at Binghamton.
────── . 1976. "Comprehension versus production in linguistic theory." *Foundations of Language* 14:525–40.
Sullivan, E., G. McCullough, and M. Stager. 1970. "A developmental study of the relation between conceptual, ego and moral development." *Child Development* 41:399–412.
Suppe, F., ed. 1977. *The Structure of Scientific Theories*. 2d ed. Urbana: University of Illinois Press.
Suppes, P. 1966. *Introduction to Logic*. New York: Van Nostrand.

Tart, C. 1975. *States of Consciousness*. New York: Dutton.
Teilhard de Chardin, P. 1959. *The Phenomenon of Man*. New York: Harper & Row.
Teleki, G. 1973. *The Predatory Behavior of Wild Chimpanzees*. Lewisburg: Bucknell University Press.
Toulmin, S., and J. Goodfield. 1961. *The Fabric of the Heavens*. New York: Harper & Row.
Trungpa, C. 1976. *The Myth of Freedom and the Way of Meditation*. Boulder, Colo.: Shambhala.
Tuggle, H., A. Townsend, and R. Riley. 1972. "Laws, systems, and research designs: A discussion of explanation in archaeology." *American Antiquity* 37:3–12.
Turiel, E. 1966. "An experimental test of the sequentiality of developmental stages in the child's moral judgments." *Journal of Personality and Social Psychology* 3:611–18.
Turnbull, C. 1972. *The Mountain People*. New York: Simon & Schuster.
────── . 1978. "Rethinking the Ik: A functional nonsocial system." In *Extinction and Survival in Human Populations*, ed. C. Laughlin and I. Brady. New York: Columbia University Press.
Tyler, S., ed. 1969. *Cognitive Anthropology*. New York: Holt, Rinehart & Winston.

Vandamme, F. 1972. *Simulation of Natural Language*. The Hague: Mouton.

Waddington, C. H. 1957. *The Strategy of the Genes.* London: Allen & Unwin.

Wallace, A. F. C. 1957. "Mazeway disintegration: The individual's perception of socio-cultural disorganization." *Human Organization* 16:23–27.

———. 1970. *Culture and Personality.* 2d ed. New York: Random House [first edition 1961].

Watson, J. 1920. *Behaviorism.* New York: W. W. Norton.

Watson, P., S. LeBlanc, and C. Redman. 1971. *Explanation in Archaeology: An Explicitly Scientific Approach.* New York: Columbia University Press.

Watzlawick, P., J. Weakland, and R. Fisch. 1974. *Change: Principles of Problem Formation and Problem Resolution.* New York: W. W. Norton.

Welsch, R. 1983. "Traditional medicine and western medical options among the Ningerum of Papua, New Guinea." In *The Anthropology of Medicine: From Culture to Method,* ed. L. Romanucci-Ross, et al. New York: Praeger.

Werner, H. 1957. "The concept of development from a comparative and organismic point of view." In *The Concept of Development,* ed. D. Harris. Minneapolis: University of Minnesota Press.

Werner, O., and J. Fenton. 1973. "Method and theory in ethnoscience or ethnoepistemology." In *A Handbook of Method in Cultural Anthropology,* ed. R. Naroll and R. Cohen. New York: Columbia University Press.

Whitehead, A. N. 1929. *Symbolism: Its Meaning and Effect.* New York: Macmillan Co.

———. 1960. *Process and Reality.* New York: W. W. Norton.

Wilber, K., ed. 1982. *The Holographic Paradigm and Other Paradoxes: Exploring the Leading Edge of Science.* Boulder, Colo.: Shambhala.

Wilson, E. O. 1975. *Sociobiology.* Cambridge, Mass.: Belknap/Harvard University Press.

Wimsatt, W. 1979. "Reduction and reductionism." In *Current Problems in Philosophy of Science,* ed. P. Asquith and H. Kyburg. East Lansing, Mich.: Philosophy of Science Association.

———. 1980a. "Randomness and perceived-randomness in evolutionary biology." *Synthese* 43:287–329.

———. 1980b. "Reductionistic research strategies and their

biases in the units of selection controversy." In *Scientific Discovery*, volume 2, *Historical and Scientific Case Studies*, ed. T. Nickles. Dordrecht, Holland: D. Reidel.

——— . 1981. "Robustness, reliability, and overdetermination." In *Scientific Inquiry and the Social Sciences*, ed. M. Brewer and B. Collins. San Francisco: Jossey-Bass.

Wittgenstein, L. 1961. *Tractatus Logico-Philosophicus*. New York: Humanities Press.

Zamenhoff, S., E. van Marthens, and L. Granel. 1971. "Prenatal cerebral development: Effect of restricted diet, reversal by growth hormone." *Science* 174:954–55.

Author Index

Ager, T., 89, 96–97
Almeder, R., 98
Arieti, S., 82, 149, 155–56
Aronson, J., 89, 96–97, 110

Bandler, R., 152, 153
Barnett, H. G., 57
Bateson, G., xviii, xx, 151, 157
Beck, A., 151
Belshaw, C., 57
Berry, J., 123
Beth, E., 33
Bharati, A., 129, 130
Binford, L., 86, 104, 119
Binford, S., 86, 104, 119
Blackburn, T., 34–35
Blalock, H. M., 92
Block, J., 56
Bloefeld, J., 129
Bohm, D., 139
Bordes, F., 119
Bovet, M., 76
Boyd, R., xvii
Brady, I., xix, 10, 25, 29, 50, 55, 56
Braithwaite, R. B., 107
Broughton, R., 145
Buckley, W., 92
Burton, A., 43

Cantril, E., 27
Chagnon, N., 10, 87
Chaney, R., 1–2
Chapple, E., 80, 85
Chenall, R., 104

Chomsky, N., 17
Coffman, T. L., 54
Cole, M., xix, 37
Conklin, H., 2
Coon, C., 80
Count, E. W., xviii, 39, 55, 56, 85, 131–32
Cove, J., 136
Crook, J., 139

d'Aquili, E., xix, 10, 12, 14, 27, 28–29, 58, 79, 80, 85, 87, 90, 93, 94, 154
Dasen, P., 42, 60, 123
Deetz, J., 119
Denham, W., 56
Donaldson, B., 119
Driver, M., 31, 40, 45–48, 54, 82
Durkheim, E., 91

Einstein, A., 73, 82, 86, 156
Eliade, M., 154
Erikson, E., 153
Evans-Wentz, W. Y., 129, 130

Feldman, C., 39, 45, 59
Fenton, J., 4, 5, 12, 16
Ferguson, M., 139
Feyerabend, P., 95
Finnegan, R., 123
Fisch, R., 153
Fishbein, H. D., 39, 85
Flavell, J., 40, 42, 43, 58
Fortes, J., 67
Frake, C., 3

177

Fritz, J., xvii, 86, 101, 104
Furth, H., 43, 44–45

Galanter, E., 131–32
Gardner, H., 138
Giere, R., 10
Gilligan, C., 50, 51
Ginsburg, H., 42
Goodfield, J., 34, 136
Granel, L., 57–58
Greeno, J., 111–12
Grinder, J., 152, 153
Grof, S., 150

Haan, N., 56
Hager, J., 40, 69
Hammel, E., 2
Hanson, N. R., xxi, 67, 95
Harvey, O., 27, 41, 45
Hastorf, S., 27
Hebb, D. O., 143
Hempel, C., xxi, 6, 88, 101–15
Hepburn, M., 38
Hill, C., 54
Hill, J., 120
Hinde, R., 92
Horton, R., 90, 123
Hufford, D., 34, 93
Hume, D., xxii, 28–29, 123–24
Hunt, D., 27, 41, 45
Hunter, I. M. L., 7
Hymes, D. H., 18

Inhelder, H., 42, 76, 139
Irons, W., 87

Jacobi, J., 149
Jarvie, I. C., xvii, 104–5, 123
Jeffery, R., 108–9
Jung, C. G., 149, 150, 152, 159

Kagan, J., 38
Kammerer, P., 69–70
Kaplan, A., 31–32
Kazdin, A., 70–71
Kearsley, R., 38

Kenigsberg, M., 70–71
Kleinbaum, D., 10
Klinger, E., 145
Koertge, N., 105
Koestler, A., 66, 69–70, 155–56
Kohlberg, L., 41, 50–51, 52
Koplowitz, H., 139
Kuhn, T. S., xxi, 61–83, 95, 124–25
Kupper, L., 10

Lancy, D., 5
Langer, J., 39, 43, 58
Laughlin, C. D., xix, 10, 14, 18, 23, 25, 27, 30, 50, 55, 56, 79, 85, 87, 90, 94, 111, 157
LeBlanc, S., 86, 101, 119
Lee, B., 39, 45, 59
Lévi-Strauss, C., 55, 59–60, 78–79, 91, 123
Levy-Bruhl, L., 123
Lomnitz, C., 67
Lomnitz, L., 67
Lounsbury, F., 4, 11
Luria, A. R., 28

McCullough, G., 50
McLean, J., 39, 45, 59
Mahoney, M. J., 70–71
Martindale, C., 94
Masterman, M., 61
Michalik, G., 58
Miller, G. A., 27, 131–32
Minton, C., 38
Missner, M., 17
Moore, T., 1–2
Morton, K., 1–2
Murry, J., 39, 45, 59

Nagel, E., xxi, 88, 101, 102
Neisser, U., 11, 14, 71–72
Newell, A., 14–16

Odum, E., 21
Opper, S., 42
Ornstein, R., 109

Author Index

Paredes, J., 38
Peirce, C. S., 98
Perls, F., 152
Petrie, W. M. F., 119
Phillips, J., 43
Piaget, J., xx, 4, 11, 14, 22, 27, 37, 38–40, 41–45, 47, 58, 59–60, 74–75, 76, 78–80, 82, 88, 90, 129, 139, 142–43, 145, 147, 148, 153, 155
Pillemer, D., 39, 45, 59
Pinxten, R., 53–54
Plog, F., xvii, 55, 86, 101, 104, 119, 120–21
Popper, K., 105, 130, 133
Pribram, K., 24, 40, 98, 131–32, 137, 139
Price-Williams, D. R., 38

Quine, W., 5, 9

Radford, J., 43
Radin, P., 123
Read, D., 119
Redman, C., 86, 101
Reichenbach, H., xvii, 87–88
Rensch, B., 94, 126
Riddington, R., 154, 156
Riley, R., 104
Rosenthal, R., 10
Rossi, I., 55
Rubinstein, R. A., xix, 10, 18, 29, 34, 50, 54, 56, 85, 93, 94, 111, 119, 148
Rudner, R., 86–87
Russell, B., 91

Sade, D., 56
Salmon, W. C., 86, 93, 109–10, 115–19, 124
Saltz, E., 14
Sanday, P., 16
Schneider, H. K., 105–6
Schneour, E., 57
Schroder, H., 31, 40, 45–48, 54, 82

Schroder, S., 41, 45
Scribner, S., xix, 37
Seligman, M., 40, 69
Shapere, D., 61
Sinclair, H., 76
Skinner, B. F., xvii, 91
Smith, M., 56
Smith, W. J., 154
Spaulding, A., 101, 119
Spradley, J., 5, 38
Stager, M., 50
Stephens, C., 21
Stewart, K., 129, 156
Straight, H. S., 6, 17, 18
Strathern, A., 5
Streufert, S., 31, 40, 45–48, 54, 82
Suefeld, P., 45
Sullivan, E., 50
Suppe, F., xvii, 5, 86, 87, 101, 107, 108

Tart, C., 33, 54, 94, 129, 130
Tax, S., 111
Teilhard de Chardin, P., 130
Teleki, G., 56
Toulmin, S., 34, 136
Townsend, A., 104
Trungpa, C., 129, 130
Tuggle, H., 104
Turiel, E., 27, 52, 147
Turnbull, C., 9–10, 56, 111
Tyler, S., 2–3, 4, 38

Vandamme, F., 17
van Marthens, E., 57–58

Waddington, C. H., 40, 78, 142
Wallace, A. F. C., 7, 12–14, 27
Watson, J., 26
Watson, P., 86, 101
Watzlawick, P., 153
Weakland, J., 153
Weingard, R., 89, 96–97
Welsch, R., 34

Werner, H., 40
Werner, O., 4, 5, 12, 16
Whitehead, A. N., 92, 98, 131–32, 146
Wilber, K., 139
Wilson, E. O., 91

Wimsatt, W., xvii, 5, 20, 33, 89, 93, 115, 133
Wittgenstein, L., 91

Zamenhoff, S., 57–58
Zelazo, P., 38

Subject Index

Abduction, 157–58. *See also* Logic
Accommodation: concept of, 41, 44–45, 129; and paradigm, 72, 79, 81
Active imagination technique, 152
Adaptability of cognitive structures, 39, 46
Adaptation. *See* Equilibration
Adaptive balance, 72, 94, 113–15, 138, 149
Adaptive intelligence, 39. *See also* Cognitive structure
Analysis, levels of. *See* Structure
Anomalies: and cognition, 27, 77, 80, 137; and model reformulation, 114; and paradigms, 65, 66, 67–73, 77
Anthropology: communication bias in, 51–55; content approach in, xix, 37–38; explanation in, 104–7, 111, 119–20; and philosophy of science, 86–88; scientific models in, 136; synchronic study in, 55–56; theory reduction in, 94–95. *See also* Cognitive anthropology
Antireductionism, 91–92
Archaeology: Hempelian models in, 104; Statistical Relevance model in, 119–21
Argument, explanation as, 103, 108–10, 113

Assimilation: concept of, 27, 41, 44, 45, 129; and paradigms, 72, 79, 81
Attribute class, 116–19

Behavior: and competence versus performance, 16–19; and paradigm, 66–67, 73–74, 77, 80; prediction of, 92; and structure, 77, 79–81, 114
Behaviorism, 69, 70
Berlin School, 107
Bias: in fieldwork, and cognitive development, 10, 51–55; and preoperational thinking, 148–51, 154–55; statistical, and exploratory versus confirmatory inquiry, 11
Biogenetic Structuralism, xix
Bracketing, 90, 114
Brain, 21, 138–39, 143
Bridge laws, 88–89, 96–97
Buddhism, 35, 159

Cargo cults, 105
Causality, 28–29
Children: cognitive development in, 74–76; moral-social reasoning in, 52
Cognitive anthropology: characterized, 1–6; evaluated, 7–20; Inductive-Statistical model in, 106–7
Cognitive boundary, 30

181

Subject Index

Cognitive development: and cognized logic, 58–60; concept of, 40–41, 78–79; and decalages, 147, 148; differing levels of, and bias in fieldwork, 51–55; differing levels of, as constraint on social action, 56–58; and explanation, 113–15; and malnutrition, 57–58; regressions in, 82, 147–48; structural complexity of, in conceptual systems theory, 46–47; implications of theories of, 51–60; overview of theories of, 41–51; and theory reduction, 89–96
Cognitive maps, 14
Cognitive structure: biological bases of, 39–40; collapse of, 50, 148–49, 151; complexity of, 39, 149–50; defined, 38–39; development of (*see* Cognitive development); dimensions of, 39; paradigms compared with, 74–78; permeability of, 30, 31; transformations of, 27–28, 38–39, 78–80
Cognized environment: and cognitive structure, 39; and cognized logic, 33; and operational environment, relationship between, 21–31, 98, 145–46, 158; as organization level, 145–46; organization levels modeled by, 29–31; and paradigm, 63; and prediction, 125–26; and scientific models, 132; self-modeling by, 25–28, 33; and sensory transformation, 79; types of activity in modeling by, 27–29
Cognized logic: normative, 34, 137–38; and operational logic, 32–35, 58, 98

Cognized self, 22–23
Competence versus performance, 7, 12, 16–18
Complexity: of cognitive structures, 39, 149–50; of cognized environments, 31; of operational environments, 31, 39, 47–50
Conceptual systems theory, 41, 45–50, 51
Concrete operational thought: and bias in fieldwork, 53; concept of, 42; and paradigms, 75–77
Confirmatory versus exploratory inquiry, 10–11
Constitution, U.S., interpretation of, 52
Consummatory acts, 13–14
Content versus process or structure, 37–38, 50, 60, 98
Conventional thought, 50
Correlation, and theory reduction, 97
Covering law models. *See* Deductive-Nomological (D-N) model; Inductive-Statistical (I-S) model
Creative intuition, 156–58
Culture and cognition paradox, 38
Culture shock, 54

Decalages, 147, 148
Deduction: Hempelian model of (*see* Deductive-Nomological [D-N] model); and induction (*see* Induction); in theory reduction, 88–89
Deductive-Nomological (D-N) model: in anthropology, 104–6, 119; critique of, 107–13, 115; outline of, 101–3
Depression, 151
Developmental transformations, 27–28

Diaphasis, 25, 30
Digestion, accommodation and assimilation in, 44–45
Dimensions, in conceptual systems theory, 46–47
Disciplinary matrix, 63, 64
Discovery versus justification, xvii, 87–88, 95
Double-bias situation, 53–54
Dreams, 145, 156

Economic Man (Schneider), 105–6
Ego, 27, 33, 146
Egocentricity, 144, 145, 148
Egyptians, and causality, 28–29
Emergent rules, 46
Empty-chair technique, 152
Enquiry concerning Human Understanding (Hume), 123
Entrepreneurship, 57
Equilibration: concept of, 27, 41, 43, 44–45; types of transformation in, 27–28. *See also* Accommodation; Assimilation
Ethnoscience. *See* Cognitive anthropology
Exemplars, 63, 64
Experiential knowledge. *See* Figurative knowing
Explanandum and explanans, 102–3. *See also* Deductive-Nomological (D-N) model
Explanation: in anthropology, 86; Hempelian models of, 101–13; Statistical Relevance model of, 115–22. *See also* Theory reduction
Exploratory versus confirmatory inquiry, 10–11

Fallibility of knowledge, 33–34, 98
Figurative knowing: advantages of, 155–58; concept of, 142–44; and logical thinking, 145–55, 158; in preoperational thought, 144–45
Formal analysis, in cognitive anthropology, 4, 10–11
Formal operational thought: and bias in fieldwork, 53; concept of, 42, 50; and logic, 58–59; and social action, 56–57
Functional invariants, 39–40, 43

Genetic epistemology of Piaget, 41–45, 51
Gestalt, 95
Gestalt therapy, 152

Habituation, in science and cognition, 81–82
Hempelian models. *See* Deductive-Nomological (D-N) model; Inductive-Statistical (I-S) model
Hypotheses, in scientific modeling, 132–34

Identity statements, in theory reduction, 96–97
Ik, 9–10, 56, 111
Images. *See* Figurative knowing
Improbable events, 109, 111, 119
Induction: concept of, 125–28; and deduction, as alternating phases of cognition, 126–28, 131–39; and deduction, as opposites, 86–87, 126, 127–28; Hempelian model of (*see* Inductive-Statistical [I-S] model); justification of, 123–25, 128–30; and paradigm, 62
Inductive-Statistical (I-S) model: in anthropology, 106–7; critique of, 107–13, 115; outline of, 101–3
Inference, explanation as, 103, 108–9, 113
Informants, use of, 52–53

Information: and cognitive structure, 39; sensory transformation of, 28–29, 79
Information-processing, and conceptual systems theory, 45, 50
Initial condition, 102, 105, 106
Instinctive knowledge, 142
Instrumental acts, 13–14
Intelligence, Piaget's concept of, 42, 45
Intracultural variation, 12, 38, 63–64
Introspection, xviii, 8, 9–10, 88
Isomorphism: of operational and cognized environments, 23–25; of operational and cognized logics, 33; of primary and secondary models, 25–27

Justification: discovery versus, xvii, 87–88, 95; of induction, 123–25, 128–30

Kepler, Johannes, and orbit of Mars, 134–36
Kinship, 107
Knowing, kinds of, 142–44

Lamarckians, 69–70
Language: and cognition, 26; competence versus performance in, 17; and structure, 4, 5
Levels of structure, 27–31, 43–50, 87; in explanation, 114–15; and paradigms, 64–67; and theory reduction, 93–94
Light, speed of, 73
Logic: and cognitive development, 58–60; operational and cognized, 32–35, 58, 98; transductive, 59–60, 145, 149, 150, 156, 157
Logical atomism, 91–92
Logical Positivism, 107–8

Logico-mathematical knowledge, as form of knowing, 142, 143

Malnutrition, 57–58
Mathematics, 33, 58, 142, 143
Minimal inclusion rule, 93
Mixed rules, 46
Models: cognitive (see Cognized environment); scientific, 131–36
Monophasia of science, 129–30
Moral development theory of Kohlberg, 41, 50–51, 52
Myth, 59–60, 155

Nematodes, 21–22, 24
Neo-Darwinians, 69–70
Nervous system: and cognized environment, 22, 24, 25, 26–27, 29–30; and instinctive knowing, 142; and perception, 72; and preoperational thought, 154
New Astronomy (Kepler), 135–36
Newtonian physics, 82
"Normal science," 62, 64–65, 68, 77
Normative cognized logic, 34, 137–38

Objectivity, in science, 130, 146–47, 148–49
Operational environment: changes in, 25, 30; and cognitive structure, 40, 41, 79, 84; and cognized environment, relationship between, 21–31, 98, 145–46, 158; complexity of, 31, 39, 47–50; external and internal, 22–23, 146–47, 148, 149–50, 158; and figurative knowing, 142–44; as organization level, 145–46; and Piaget's developmental stages, 42; and scientific models, 131–33

Subject Index

Operational logic, 32–35, 58, 98
Operations, in genetic epistemology, 42
Organization, as functional invariant, 43–44. *See also* Structure

Paradigms: cognition compared with, 73–78, 80–82; cognized logics as, 34–35; concept of, 61–63; and figurative knowing, 154–56; levels of, 63–64; and perception, 67–76; and phases of science, 64–67; and ritual, 64, 79–80, 82–84
Parsimony, 6, 8–9
Perception: and figurative knowing, 143; and paradigms, 66, 67–76. *See also* "Seeing"
Performance versus competence, 7, 12, 16–18
Permeability of cognitive structures, 30, 31
Phenomenology, 129–30, 159
Philosophy of science, as discipline, xvi–xviii, 86–88
Postconventional thought, 50, 52
Post hoc fallacy, 134
Preconventional thought, 50, 52
Prediction, 125. *See also* Theory
Preoperational thought: concept of, 42, 58, 59–60; figurative thought as part of, 144–45, 150, 158
Primary equivalence structure, 13–14
"Primitive" thought, 60, 123
"Principles of reason," 59
Process versus content, 37–38, 50, 60
Projection, 144–45, 146, 148, 158
Psychoanalysis, 146
Psychological reality, 7–11
Psychology: paradigm in, 69; process approach in, 37–38

Psychopathology, and cognitive functioning, 82, 149, 151–52
Puzzle-solving activity, 62, 64

Rationality principle, 105
Reality: as modeled by brain, 21–22; psychological, and cognitive theories, 7–11
Reconstructed logic. *See* Cognized logic
Reference classes, 115–17, 119. *See also* Statistical Relevance (S-R) model
Regressions in cognitive functioning, 82, 147–48
Revolution in Anthropology, The (Jarvie), 104–5
Ritual: and nonlogical thought, 153–54, 156; and paradigm, 64, 79–80, 82–84
Rules, in conceptual systems theory, 46–47

Schema, 72
Schizophrenia, 149
Sciencing, defined, xviii–xix, 114
Scientific induction. *See* Induction
Scientific method, *xvi*, 27
Scientific models, 131–36
Scientific training, 35, 64, 83
Screening-off relation, 118, 121–22
Secondary equivalence structure, 13–14
"Seeing": in Buddhism, 35; in cognitive development, 75–76; by scientists, 63, 64, 66, 67–73, 77; and "seeing that," 95; "thinking" compared with, 143. *See also* Perception
Self-regulation, 38–39
Sensory cortex, 143
Sensory motor intelligence, 42
Sensory transformations, 28–29, 79

Subject Index

Shamanism, 154, 159
Sharedness of cognitive systems, 7, 11–16
Simplicity. *See* Parsimony
Situational logic, 105
So, 56
Social intervention, 57–58
Societal infrastructure, 29
Sociological fallacy, 92
Stability of cognitive structures, 39
State dependence, 130
Statistical generalizations, in Inductive-Statistical model, 102–3, 111–12
Statistical Relevance (S-R) model: in anthropology, 119–22; outline of, 115–19
Stress, and cognitive functioning, 48–50, 53–56, 57, 82
Structural elaboration model, 89–96
Structure: deep, 16–17, 85; linguistic, 4, 5; and minimal inclusion rule, 93; modeled by cognitive environment, 29–31; and ontogenetic growth, 114–15; Piaget's concept of, 3–5, 37–39, 43–44; relationship between levels of, 85; as subject of cognitive anthropology, 3–6, 16–17; surface, 16–17, 29–30, 80–81, 85. *See also* Cognitive structure
Suprainformant, 12

Surface transformations, 28, 78–79
Systemic collapse, 50, 148–49, 151

Taxonomic analysis, 5
Theories: received view of, 107–8; reduction of (*see* Theory reduction); as schemata, 114
Theory reduction: by incorporation, 96–99; received view of, 88–89; structural elaboration model of, 89–96. *See also* Explanation
Thorndike-Lorge U-curve function, 31, 82, 148
Tibetan Book of the Dead, 150
"Trained intuition," 34–35
Transduction (transductive logic), 59–60, 145, 149, 150, 156, 157
Transformation: as component of structure, 38–39; equilibrating, 27–28, 78–79; rules of, 4
Treatise of Human Nature (Hume), 123
Turiel effect, 52, 54

Universal laws, and explanation, 102, 103, 110

Vienna Circle, 107

Wholeness, 38
World view, 63. *See also* Paradigms